ELECTRONIC WARFARE

PRINCIPLES OF ELECTRONIC WARFARE

*BY MEMBERS OF THE TECHNICAL STAFF OF CONVAIR,
A DIVISION OF GENERAL DYNAMICS CORPORATION*

Robert J. Schlesinger
CONVAIR GENERAL OFFICES

WITH THE COLLABORATION OF

Kirk Abbey
SAN DIEGO DIVISION

Richard W. Ehrhorn
POMONA DIVISION

Kenneth J. Friedenthal
GENERAL OFFICES

Stanley H. Logue
SAN DIEGO DIVISION

P.O. Box 867, Los Altos, California 94022

It is not pleasant to think that warfare is fundamental to the nature of man, and indeed it may not necessarily be so. However, through all recorded history we find this unfortunate trend. Perhaps some day a discussion of the problems taken up in this book will be pointless; i.e., wars will no longer exist. Let us hope, for that reason, that the material herein will then no longer be required.

In that light, this book is dedicated to its own obsolescence.

© 1961 by PRENTICE-HALL, Inc., Englewood Cliffs, New Jersey.

All rights reserved. No part of this book may be reproduced in any form, by mimeograph or any other means, without permission in writing from the publisher. Printed in the United States of America. Library of Congress Catalog Card Number 61-15515.
70923—C
ISBN 0-932146-01-5
This edition has been published with permission of the original publisher, Prentice-Hall, Inc., Englewood Cliffs, N.J.

PREFACE

During World War II, the rapid growth of military electronics had its start. Today, and in the foreseeable future, large-scale weapon systems depend and will depend on electronic technology.

The importance of these electronic weapon systems to the overall concept of warfare is often obscured in detailed discussions of specific systems. For this reason, a broad understanding of the principles involved is essential. It is significant that these principles involve both engineering sciences and military strategy. In electronic warfare the terms "radiation" and "detection" must be considered in the same light as "offense" and "defense" are in strategic and tactical warfare. To be sure, the engineering aspects of the problem must be given thorough analytical treatment; however, operational questions must also be considered if a broad appreciation of the principles of electronic warfare is to be achieved. The treatment presented here is intended to achieve a balance between these technical and tactical aspects of the problem. Therefore, it is hoped that both those concerned with deployment and tactics, and the professional engineer as well, will find many points of interest here.

Some aspects of the Electronic Countermeasures (ECM) problem have intentionally been omitted because security requirements impose an important constraint on the selection and treatment of topics.

There is no discussion of specific equipment in this text. There are two reasons for this intentional omission. First, a discussion of operational equipment is not our objective; second, such a treatment does not represent a fundamental approach to the problem. A sincere attempt has been made to bring reasonable generality to all mathematical analysis.

The opening chapter is intended to indicate the range of the subject matter and to point up typical situations representative of "electronic warfare." Chapter 2 represents an attempt to relate the broad concepts of techniques and tactics and also to provide a general discussion of some classic ECM methods.

In Chapter 3, we provide a basic introduction to the mathematics associated with probability and then apply this to a discussion of noise theory, which is inherent to many parts of the problem. In Chapter 4 we develop the theory of electronic intelligence and then discuss some basic problems of reconnaissance systems. It is intended here to establish the relationships between reconnaissance inputs and ECM tactics. In Chapter 5 we carry out a fundamental analysis of radar systems, with emphasis on parameters important to ECM and ECCM techniques. In Chapter 6 we define the role of the antenna as the transducer between the electronic system and the environment, and again point out factors critical to the electronic warfare problem. In Chapter 7 we provide an operations analysis approach to the problem of establishing an effective ECM environment on a mission profile with typical constraints. In Chapter 8 we extend present thinking into the space era and discuss some of the ramifications of electronic warfare as influenced by space flight.

Although this book is the composite effort of all contributors, Mr. Ehrhorn is primarily responsible for Chapter 3, Mr. Logue for Chapter 5, Mr. Abbey for Chapter 6, and Mr. Friedenthal and Mr. Schlesinger for Chapters 1, 2, 4, 7, and 8.

Robert J. Schlesinger

ADDENDUM TO PREFACE

Principles of Electronic Warfare was originally published in 1961 as the first book in open literature that provided a systems approach to Electronic Warfare. It is now reprinted in its original, unabridged form because it is as valuable today as when first published. To those concerned with either operational tactics or basic engineering, the book presents an understandable overview of the subject.

January 1979   *Peninsula Publishing*
Los Altos, California

CONTENTS

CHAPTER ONE
The Scope of the Problem   1

Some elements of the problem, 1; Technical supremacy and the secret weapon, 3; Dynamic development of an ECM tactic, 5; ECM equipment—one element of the total system, 7.

CHAPTER TWO
Technique and Tactics   10

Tactical considerations, 10; Noise jamming, 13; Deception jammers, 14; Other radar confusion devices, 15; Typical air-combat analysis in an ECM environment, 17; Tactics—an example, 21.

CHAPTER THREE
Noise, Probability, and Information Recovery   26

Relationship of "noise" to an information-carrying system, 26; Elementary treatment of noise theory, 26; Elementary probability and statistics, 31; Recovery of signal information from a noise environment, 35; Detection and correlation, 38; Signal detection in the presence of electronic countermeasures (ECM), 42.

CHAPTER FOUR
Electronic Reconnaissance   51

The purpose of electronic reconnaissance, 51; Electronic intelligence, 52; A. Operational information, 52; B. Tech-

viii CONTENTS

CHAPTER FOUR—Continued

nical data, 54; Probability of detecting, probability of intercepting, 56; Searching modes, 57; Probability of overflight, 60; Overflight resolution, 62; "Horizontal" resolution, 65; Intercept correlation, 71; Generalized "sensor" characteristics, 73; Receiver modes, 76.

CHAPTER FIVE
Radar Considerations — 81

The approach to radar countermeasures, 81; Pulse radars, 82; Pulse-radar operation, 82; Optimum IF amplifier band width, 86; Pulse detection and integration, 89; Pulse-radar range, 92; CW radars, 94; Doppler-filter band width limitations, 96; CW radar range, 102; Pulse-Doppler radars, 103; Moving target indication, 109; Target tracking systems, 111; Angle tracking, 111; Range tracking, 114; Velocity tracking, 116; General radar concepts, 116; The effects of ECM on radars, 123; Radar countermeasures techniques, 126.

CHAPTER SIX
The Role of Antennas in Electronic Warfare — 132

Antenna descriptive parameters, 133; Directivity and gain, 136; Antenna side lobes, 137; Band widths, 137; Propagation, 139; Propagation over a plain earth, 139; Propagation over the curved earth, 140; Propagation in the standard atmosphere, 141; Propagation in a nonstandard atmosphere, 143; Attenuation due to absorption and scattering, 144; Antenna applications, 144; Tracking antennas, 144; Manual tracking, 146; Multiple-beam tracking, 146; Conical scan, 146; Monopulse tracking, 147; Search antennas, 149; Effect of antenna characteristics on intercept probability, 149; Design of large ground-based antennas, 151; Antennas of other types, 153.

CHAPTER SEVEN
Optimization—Constraints and Incomplete Information — 154

Objectives of an ECM system, 154; Optimization of ECM equipment characteristics, 155; Definition of the effectiveness function, 163; Optimum distribution of available payload between weapons and ECM, 170.

CHAPTER EIGHT
Some Aspects of Electronic Warfare in the Space Era 176

Introduction, 176; Some elements of the space environment, 178; Space missions—communications and reconnaissance, 187; Flush antennas, 195; Inflatable and unfolding antennas, 196; Polarization, 196; Antenna materials for space systems, 197; Infrared applications, 197; Some effects of space operations on radar and radar countermeasures, 198.

Bibliography, 202

Index, 205

I

THE SCOPE OF THE PROBLEM

SOME ELEMENTS OF THE PROBLEM

It is our goal in this book to set down some of the existing theories, explore the current philosophies, and define the present problems connected with electronic warfare.

One of the first problems to consider is a definition of electronic warfare that will be acceptable in the situations to be discussed. In the context of this book, the interaction between two or more communication systems for the purpose of intentional interference will represent electronic warfare. A communication system is here understood to be any electronic device that radiates and/or receives information.

The problems concerned with when, where, and how to generate this electronic interference and, on the other hand, the action to be taken to counter its detrimental effects are of fundamental importance to electronic warfare. The former problem is generally referred to as Electronic Counter-Measures (ECM) and the latter as Electronic Counter-Counter Measures (ECCM).

It is the end purpose of ECM to interfere with the successful operation of an opponent's weapon system; particularly weapon systems that might be used to destroy the vehicle (for example, an aircraft) on which the ECM equipment is carried.

Since the jamming of radar, communication, and missile guidance systems tends to accomplish this end purpose, these are the areas in which ECM has found its widest application.

It is natural to expect that as weapon systems came to place more reliance on the use of electromagnetic radiation as a connecting link among their elements, the weakness inherent to such a link would be exploited. Unfortunately for the user, one of the weaknesses of these systems lies in the fact

2   SCOPE OF THE PROBLEM

that the radiation of Radio Frequency (*RF*) energy, in most of its applications, is not secure from detection and interference. To be sure, considerable effort is expended in some applications, notably communications, to provide some degree of security. However, since all weapon systems of the type considered here must transmit information in order to be useful, their security becomes a function of the detector used to intercept the signal. If the enemy has exactly the right type of receiver for the signal characteristics being transmitted, security is extremely difficult, if not impossible.

The subject of security is considered here because of its basic importance to the intelligent application of ECM methods. If the presence of a hostile signal cannot be detected, no jamming action can be initiated wisely. After detection has been accomplished, it is necessary to establish the information content of the signal to determine if jamming is required. Typical techniques of preventing these two steps from being carried out include high-speed transmissions and pseudo-noise generation with correlation to impede detection, and coded modulation methods to prevent information analysis. Many other techniques are used, some of which will be discussed in later sections of this text.

We have introduced two of the primary requirements for the employment of ECM: (1) it is necessary to detect the radiation from an enemy system; (2) it must be established that it is desirable to jam the signal detected.

It is not always in the best interest of the mission to jam every signal being emitted within enemy territory. For example, if an incoming bomber raid detected radiation, but did not analyze the signal, ECM transmitters might be turned on to jam a local UHF television station. This would hinder, more than aid, the success of the raid. Once the jamming transmitters are turned on, a public announcement is given that hostile aircraft are in the area. Because of the "beaconing" effect* of these transmitters, the area alerted is much larger than that covered by the surveillance radars alone.

In the foregoing discussion two important steps were defined that are necessary to the evaluation of the electronic environment with respect to the presence of hostile signals. This evaluation has up to this point depended only on the type of monitoring equipment carried aboard the penetrating aircraft and is therefore basically a reconnaissance function. If reconnaissance is the primary purpose of the flight it is not generally necessary to carry jamming transmitters.† A record of the signals intercepted can be stored on tape and/or film for use at a later time. (The application of reconnaissance information will be covered in the fourth chapter.) However, if the mission of the raid is to reach a specific target the use of active ECM techniques may be required to confuse and jam enemy air-defense systems.

When sufficient information has been received to determine that jamming tactics should indeed be employed, a second set of problems arises. These

---

* See page 6 for beaconing effect.

† In some special cases jamming transmitters are carried on reconnaissance flights. See page 55, Chapter 4.

problems concern the actual ECM tactical considerations. For example, some of the questions that must be answered include: What mode of jamming should be used? How long should it be used? How many of the aircraft should use it? Exactly when should the jamming transmitters be turned on? To answer these questions, something must be known about the operational constraints imposed by the raid configuration, its mission, and the characteristics of the ECM equipment being carried.

It can be seen that the introduction of so many variable factors makes a general solution to the ECM tactics problem extremely difficult. However, if a specific raid is considered and previous reconnaissance data have given some details of the hostile electronic environment to be encountered in the target area, optimum ECM tactics may be established. Unfortunately, this information can seldom be provided in the detail desired.

The Strategic Air Command (SAC) force required to penetrate deep into enemy territory would like very much to have a general tactic to employ in the face of an unknown electronic environment. This is a basic problem that has been of major concern to both military and civilian planners engaged in establishing electronic-warfare strategies. Before further consideration of this problem is undertaken, one additional complication and some illustrative examples will be discussed.

TECHNICAL SUPREMACY AND THE SECRET WEAPON

In many studies it has been shown that "technical supremacy" is a critical factor. Should SAC bombers encounter an air-defense system using tracking and missile-guidance radars operating at a frequency of 40 kmcs, when the highest-frequency ECM equipment they carried was 30 kmcs, the results could be fatal! Generally, since the monitoring receivers do not search higher in frequency than their companion jamming transmitters, the first indication the raid would have of their being illuminated by the enemy defense system would be the explosion of surface-to-air or air-to-air missiles. In this case, even if they had employed their ECM transmitters blindly, no amount of jamming would have reduced the effectiveness of the enemy air-defense system. The SAC bombers were not technically equipped to defeat the threat they encountered in this example. However, the correct employment of chaff at the proper time may have been of some aid. It is not uncommon to find chaff and active ECM used simultaneously in certain cases. For deep penetration raids, requiring many hours of flying time over enemy territory, it is impossible to carry enough chaff for continuous dispersal. Therefore, some operational information to establish the correct time to disperse the chaff is still required.

Perhaps the most interesting case of technical supremacy is the now classical example of submarine searching carried on during World War II.*

* Morse and Kimball, *Methods of Operations Research*, Wiley, New York, 1951.

4   SCOPE OF THE PROBLEM

In early 1942, the RAF Coastal Command used L-band radar as an aid for locating German U-boats recharging batteries on the surface. The overall effectiveness of the RAF in this task was quite good until the U-boats began using L-band search receivers. These receivers allowed the submarine to hear transmitted radar signals at a range greater than that over which the radar echo could effectively be returned. The U-boat therefore had time to crash-dive before actually being sighted by the searching aircraft. In turn, the general effectiveness of the RAF anti-submarine effort decreased. The Allies, realizing what had happened, installed new S-band search radars aboard their aircraft during early 1943. As a result of the effectiveness of new equipment the intercept rate rose sharply. German submarines sitting on the surface, listening to L-band search receivers, became vulnerable targets for S-band radar directed aircraft.

As the U-boat sinkings increased, the Germans tried frantically to determine the method of detection the Allies were using. Since reports from surviving submarines stated that no radiation had been heard in their L-band search receivers prior to the attack, it was thought that perhaps an infrared detection device of some type was being employed. Considerable effort was spent in an attempt to combat a non-existing infrared threat. U-boat activity was greatly reduced by the time the German High Command realized that a new high-frequency radar (S-band) was in use.

This is an interesting example of a weapon (L-band radar), a counter measure (L-band search receiver), and an improvement (S-band radar) providing a clear margin of technical supremacy.

There is another point to be considered. To be sure, the use of S-band radar employing magnetrons and extending the useable frequency by a factor of ten provided a definite advantage. However, had the Germans had information as to what was being used, the time lag until they were able to develop an effective S-band search receiver would have been greatly reduced. An added advantage was gained by the Allies because of the Germans' lack of information. It is obvious then that the enemy's lack of information is the basic requirement of the so-called "secret weapon."

This point is mentioned here because illustrated in this example is one of the important roles of electronic reconnaissance. Had the Germans been conducting an extensive reconnaissance program at the time, it is probable that they would have intercepted S-band signals from magnetron oscillators in the development and testing stages during flights over England. The development of the magnetron was, of course, the crux of the problem of generating high power for 10 cm radar, but simple crystal receivers for reconnaissance purposes were indeed available, if the Germans had cared to use them in this application. Sensitivity is, of course, not necessary for intercepting high-power sources. Therefore, special requirements of a reconnaissance system include being general enough to intercept the unexpected and providing intelligence inputs for an advanced ECCM program. Other relationships between ECM, ECCM, and reconnaissance will also be taken up in the fourth chapter.

DYNAMIC DEVELOPMENT OF AN ECM TACTIC

It was stated, prior to these examples, that it is difficult to specify a general ECM tactic to be employed on any and all raids. When intelligence information is complete and exact on the defenses of any given area or target (the precise types and location of all air-defense systems are known), a detailed

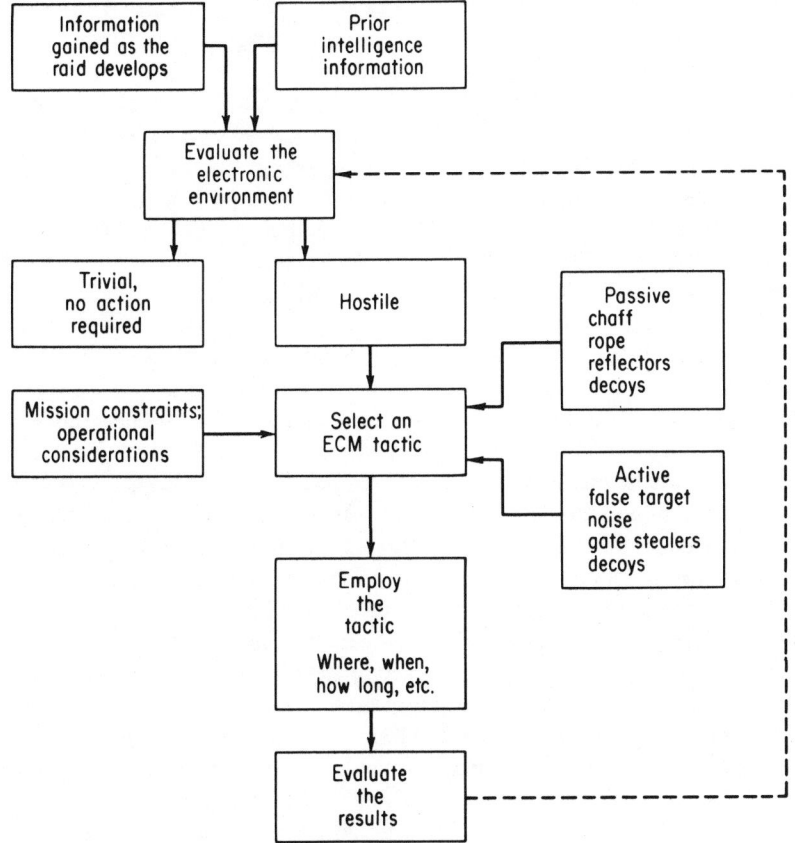

Figure 1-1. *Dynamic development of ECM tactics.*

ECM tactic can be prescribed prior to raid penetration. Unfortunately, information is seldom that complete. Although a general tactic cannot be defined beforehand, a general approach to establishing the optimum tactic as the raid progresses can be defined. Figure 1–1 shows the steps in the development of the problem.

6 SCOPE OF THE PROBLEM

Generally, some information obtained from reconnaissance or other forms of intelligence analysis is on hand. It is this information that must determine the ECM equipment to be carried on the mission. Since the weight of the ECM equipment will displace bomb loading (or equivalent firepower in the case of ground forces), its selection should be carefully considered. Obviously, once the mission has started, if incorrect or insufficient jamming units are being carried, little can be done to remedy the situation. As the raid penetrates into unfriendly territory, continuous sampling of the electronic environment is carried on. Search receivers capable of controlling jamming transmitters, as well as various types of chaff-dispensing systems, continuously evaluate intercepted signals. The results may be presented in various display forms to a member of the crew or may be programmed to allow automatic operation of the aircraft-defense subsystem. If the latter operation is selected, a manual override is generally allowed for. The outputs from this evaluation can be considered trivial, in which case no action is required; or hostile, in which case some defensive action is indeed required.

The decision as to the defensive tactic to be employed is a complex function. It is at this point that it is determined whether or not the correct equipment was selected earlier. Optimal selection of initial equipment, based on the information available prior to the raid, will be discussed analytically in Chapter 7. Selection of the best ECM tactic to use is obviously restricted by the equipment on hand. Mission restrictions also affect the decision. In the case of aircraft, which are of primary concern here, the number involved in the flight and the altitude (high- or low-level penetration) has a bearing on the problem. In view of these conditions, the ECM tactic is selected when the presence of a hostile environment is established. Some of the details involved in this selection will be taken up later. Once the tactic to use is chosen (barrage jamming, spot jamming, chaff, and so on),* a decision as to when to start using it and for how long is required. In short, operational considerations of the tactic selected must be established. Generally, an approaching aircraft carrying a searching receiver can hear the signal emitted from a surveillance radar before a strong enough echo is reflected for the radar receiver to detect the aircraft. This fact also provides time, in the order of minutes, between the determination that a hostile environment is being approached and the decision that defensive action must be initiated. Should the jamming transmitters be turned on too soon, since their radiated power is much greater than the returned radar echo, the jamming signal could be detected by the radar operator before he actually received a return pip on his scope. This early action would effectively extend the surveillance range of the radar and is referred to as "beaconing." Obviously, this condition is undesirable from the point of view of the penetrating aircraft, since it reduces the time during which it can escape detection. This, therefore, is one of the important operational considerations; i.e., when to start (and stop) jamming.

The last phase to consider is the evaluation of the ECM effectiveness. In

---
* A description of these techniques will be given in Chapter 2.

the grossest sense the survival of the aircraft is the determining factor. However, as interceptor aircraft and ground-to-air missiles are brought into action against the raid, a finer evaluation of a given ECM effort can be made. One of the methods that provides this evaluation is the technique known as "look-through." In this scheme, the jamming transmitter is turned off for a period of milliseconds and the companion search receiver determines how the fighter, or missile, radar is operating. If this pause indicates that the jamming has been effective, it is continued; if not, it allows the operator to select a different jamming mode. This technique functions automatically; the off-time is kept very short and the radars cannot lock on during this period. The results are displayed to the operator so that a manual change of jamming can be accomplished when desired.

The cycle as shown in Fig. 1-1 is thus completed; it represents the sequence of major events in the application of ECM. The foregoing discussion served to introduce a logical development for the application of ECM. In summary, it defined four basic steps:

1. Evaluate the electronic environment.
2. Select an ECM tactic.
3. Employ the tactic.
4. Evaluate the results.

ECM EQUIPMENT—ONE ELEMENT OF THE TOTAL SYSTEM

The above points cover technical and operational aspects of the ECM problem under the assumption that jamming will be used. However, the question might reasonably be asked, "By what criterion is it determined that ECM should be used at all?"

To be sure, jamming serves no useful end in itself. In fact, it is highly unlikely that a flight of aircraft, all fully loaded with nothing but jamming equipment, would be directed to fly into enemy territory in time of war. On the other hand, perhaps it is equally unwise to ignore ECM and carry only bombs. Since the payload an aircraft can carry is fixed for a given range, some understanding must be established in determining the ratio of ECM-equipment weight to bomb-load weight.

It was stated early in this chapter that the end purpose of ECM is to interfere with the successful operation of an opponent's air-defense system. If this function is successful, the aircraft will deliver its bombs; if not, the bomber may be destroyed. The effectiveness of an air-defense weapon system is defined as its probability of killing (P_k) the bomber. The proper use of ECM against the defense system can substantially reduce this P_k. As discussed earlier, the more ECM equipment carried (and hence the more weight devoted to this requirement), the more likely that effective jamming can be employed. The curve in Fig. 1-2 shows the reduction of P_k for a hypothetical

8 SCOPE OF THE PROBLEM

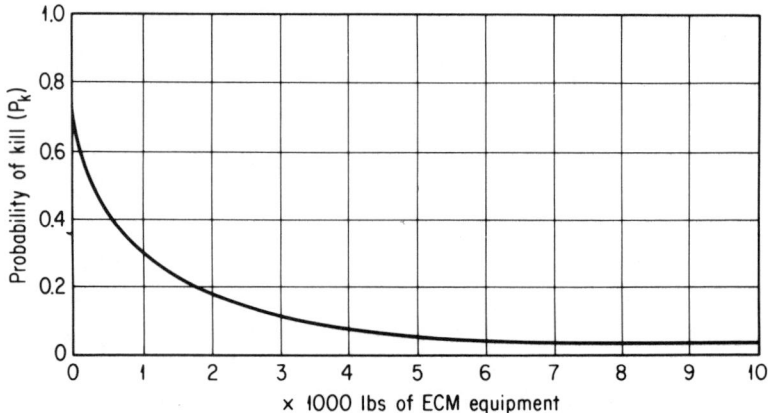

Figure 1-2. *Reduction of P_k in the presence of increased ECM as related to gross equipment weight available.*

defense weapon as a function of the ECM equipment carried. In practice, these curves can be calculated for a given system (Nike-Hercules for example) as a function of various types and levels of jamming. As more jamming is used, the probability of survival (P_s) of the aircraft increases, where $P_s = 1 - P_k$.

The curve in Fig. 1–3 shows a typical relation between the percentage of

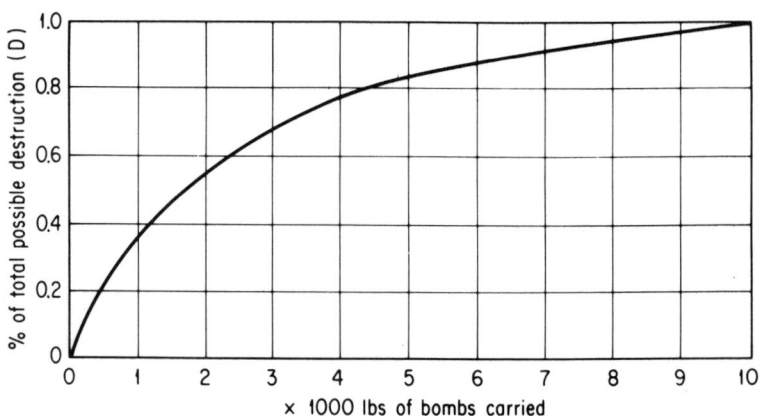

Figure 1-3. *Bomb destruction possible vs. gross bomb weight available.*

damage (D) that a bomber can deliver and the weight of bombs carried. Defining an arbitrary figure of merit:

$$Q = (1 - P_k) \times D \qquad (1\text{-}1)$$

provides a measure of the combination (ECM-weight plus bomb-weight) effectiveness. The correct ratio of ECM weight to bomb weight will occur, for a given mission, at a maximum Q. For the example data given this maximum is plotted in Fig. 1–4. In the graphs shown, a total payload weight of 10,000 pounds is assumed. This weight must be divided by some ratio between ECM equipment and bombs. In the example, Q maximum occurs at a ratio of pounds of equipment/pounds of bombs carried equal to 0.5. This is equivalent to approximately 3300 pounds of ECM equipment and 6700

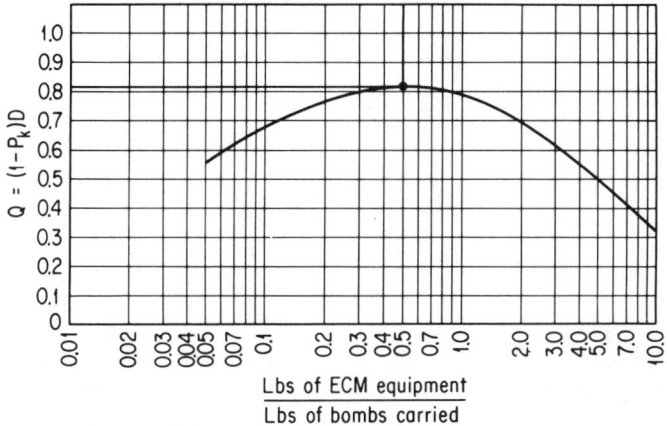

Figure 1-4. *System figure of merit, Q, vs. equipment-weight ratios.*

pounds of bombs. The result is simple to interpret. Namely, the probability of delivering some bombs by using ECM is greater than the probability of delivering a full load of bombs without the aid of ECM. Hence, it is advantageous to the mission objective to employ the correct amount of jamming tactics in a hostile electronic environment.

In this chapter we have attempted to cover some of the general aspects of the electronic-warfare problem. A rather "broad-brush" treatment of the subject was used intentionally, so that the total scope of the problem could be appreciated. The following chapters will be devoted to a more detailed discussion of the specific elements.

2

TECHNIQUE AND TACTICS

It is difficult to appreciate the significance of a detailed system analysis without a comprehensive understanding of the overall problem. It is therefore our intent to present in this chapter the general philosophies associated with electronic countermeasures activities so the discussion carried on in later chapters may be considered in its proper context.

As might be expected, there is no one tactic or system that serves for all situations. Hence, any combination selected will represent a trade-off between such factors as weight, volume, and system effectiveness. The intelligent selection of a mode of operation to fit a specific situation requires a general understanding of the relative advantages and disadvantages of possible alternatives. We shall here attempt to provide this insight through a discussion of the relative merits of active versus passive jammer modes, and noise jammers versus deception jammers. Consideration will also be given to the alternatives possible in the employment and tactics that can be developed by the use of operational analysis.

TACTICAL CONSIDERATIONS

For the purpose of this discussion it is convenient to realize that systems employed in the accomplishment of warfare consist of combinations of three elements: data-gathering elements, decision elements, and action elements. These elements are joined by communication links of various types. A set of these elements together with their associated communication links forms a functional system.

The data-gathering element (e.g., radar, sonar, radio receiver, eyes, ears) translates events that occur external to the system into a language meaningful to the decision-making element. The decision-making element translates the information received into the selection of a course of action. The action element in turn responds to this selected mode of action. Examples of the decision element may vary from simple overload relays and automatic volume control (AVC) circuits in a radio to a computer or a field commander. Action elements might include motors, ailerons, transmitters, missiles, and aircraft.

Although these definitions may seem somewhat contrived at first glance, a moment's reflection will indicate a reasonable degree of generality and validity in them. As an illustration of the definitions, consider an air-defense system. The fire-control radar (data-gathering element) tracks the incoming target. Its course, velocity, and optimum point of intercept are established by the fire-control computer (decision element) and at the proper instant a fire-control command is transmitted to the missile battery (action element). The same series of events can be studied on a much smaller scale. Consider, for example, a barometric fuze. The relative atmospheric pressure is sensed (data gathering) and at some set reference point (decision) the detonation of the weapon is activated. Both of these "systems" carry through the three steps of the necessary process for logical action.

Two of the aspects of the system operation as described above are pertinent to the use of ECM. First, for a system to perform properly the operation of each element must be accomplished satisfactorily; second, all elements do not necessarily function simultaneously, nor do they require similar information from the environment. These considerations imply that the ECM user may choose an ECM tactic to inhibit the functioning of any element, and if he is rational he will choose the weakest element. However, it also implies that the system may be devised with several parallel elements to perform a single function. Thus, to prevent the function of a particular element from being performed, all parallel elements must be confounded. Again, if the system user is rational he will seek to parallel the most vulnerable element in the system. It is this interplay of seeking weak spots and seeking to parallel or strengthen the weak spots that leads to the constant ECM-ECCM battle. As this battle proceeds a point may be reached at which the expense of attempting to inhibit the functioning of a system by ECM equals the cost of actually destroying an element of the system. At this point the ECM-ECCM battle gives way to more conventional means of warfare. For example, as early-warning radars become more and more effective a point is reached where an incoming bomber finds it cheaper to carry a radar homing missile to destroy the radar than to carry the great quantity of high-powered ECM equipment that would be necessary to jam it adequately. However, since the number of early-warning radars in operation represents parallel operation of this element of the system, the number of missiles needed to destroy them would be quite large. Jamming is still, for that reason, an attractive mode of operation.

Consider as a model for discussion a typical air-defense system of the type used by both the United States and the U.S.S.R. The three elements of interest consist of search and tracking radars, direction centers and computers, and interceptors and missile squadrons. Each phase of the system offers in varying degrees to a penetrating bomber an opportunity to jam the weapons operation. The direction centers are reasonably free from direct jamming and hence represent no point of interest here. During the initial part of a penetration the intruding bomber is under surveillance from powerful long-range ground radars. These radars do not in themselves represent a threat to the destruction of the bomber. It is of course the interceptor (manned or unmanned) that is being directed by them via a radio data link that poses the threat. Therefore, rather than engage these powerful ground radars in a "jamming duel" it is often more desirable to avoid detection as long as possible. One method of accomplishing such a tactic is to approach the target area at as low an altitude as possible, making use of terrain features and the earth's curvature to shield the bomber from radar detection. Unfortunately for the bomber, this is a fuel-consuming tactic (imposing a reduced range capability or a lower bomb-carrying capability), since modern high-performance engines do not function efficiently at low altitudes. If the defense radar is located too close to the target area the bomber may choose to approach at an optimum height to just beyond the maximum expected detection range and then drop to a low protective altitude for the final run at the target. In this case it is willing to expend the extra fuel for a relatively short distance in the hope of escaping early detection. However, when the radars are located at many points, extending a long distance from the target area, the low-altitude attack may prove too costly in fuel consumption. From this short discussion one can readily understand the importance of our Distance Early Warning (DEW) radar line as a method of providing early detection or forcing our enemy into an unfavorable attack posture.

The final interceptor attack (action element) on a bomber is a particularly critical period in the battle. Modern high-performance interceptors can be expected to make only one, or at the most two, passes at their target before fuel is expended to the point at which they must return to base. If jamming can be successfully employed for a short time during this critical period of the bomber penetration, the interceptor attack may be forced to abort.

A complete program to establish the probability of a successful terminal attack and kill by an interceptor of a bomber employing jamming is a matter of extremely complex and lengthy mathematical analysis. When consideration of all variable parameters is included, a computer program is required to investigate the whole general case, taking into account all modes of attack and all airborne weapons. However, since ground radars in themselves do not destroy attacking bombers, it is this terminal interceptor phase that is of major interest. Further, it is not just attacking bombers that are of interest here, but rather attacking bombers employing ECM. For this reason it will now be necessary to introduce the effects of various jamming systems, after which a simple attack-phase analysis in the presence of jamming can be undertaken.

NOISE JAMMING

The jamming philosophy selected can be divided into two major categories. The first of these will be referred to as the "brute-force" methods. Noise jammers of all types fall into this group. The principle here is simply to "outshout" the radar return. It is shown in Chapter 3 how the increase in noise tends to mask the desired signal. Thus, if sufficient noise can be induced in the radar-receiver band-pass by the jamming transmitter, the radar operator will be unable to distinguish the target echo.

A given radar set is generally capable of tuning over a broad band. For example, assume an X-band radar can operate on any selected channel between 9000 to 10,000 mcs. Each channel is in the order of 10 mcs wide. Hence, 100 possible channels are available to the radar operator. To employ noise jamming effectively, it is necessary to generate a given "noise-power density" on the radar channel. The measurement of this jamming function is then watts/megacycle.

The first of the noise-jamming techniques to be considered is known as *spot jamming*. In this mode the jamming-transmitter band width is about equal to the radar-channel band width (10 mcs). Under these conditions, if a 100 w jammer is tuned to the same channel (spot) being used by the radar, a jamming-power density of 100w/10 mcs = 10 w/mc results. Since a normal radar target echo is in the order of a fraction of a watt it is easy to understand how this level of jamming noise obscures the target. Although the noise-power density resulting from spot jamming is high, there is one major disadvantage to this method. Present-day radar can very quickly tune to a new channel anywhere within its operating range.

This flexibility in turn requires that the jammer search for the new radar frequency and shift its jamming frequency accordingly. Although the jammer can also retune to new frequencies very quickly, it might be possible for the radar to get a sufficient number of unjammed signal returns to track the target.

This effect can be overcome if the jammer spreads its power simultaneously over the entire radar tuning range. This mode of operation is known as *broad-band barrage jamming*. In the example discussed above for an X-band radar and a 100 w jammer, the power must be spread between 9000 and 10,000 mcs. It can be seen that the resulting power density is 100 w/1000 mcs = 0.1 w/mcs. This level is now constant over the entire radar band but is somewhat reduced in amplitude compared with the spot-jamming mode. The broad-band barrage jammer will require an average power of 10,000 w to equal the 100 w spot jammer in power density generated within the radar channel. A compromise approach between the two is made possible by the use of a *swept-spot jammer*. In this system the jammer frequency is swept at a very high rate across the range of frequencies to be jammed (9000 to 10,000 mcs). This method allows the high-power density of the spot jammer to appear for a short period on all of the possible radar channels. By careful selection of the sweep rate for the jammer it is possible to inhibit recovery of the radar receiver during the time required for the jammer to cover the entire band. However, if the radar-tuning range is very wide, or if special circuits known as "fixes" are added to the radar, this technique loses its effectiveness. Of the three brute-force noise methods, barrage jamming is the most reliable if sufficient power can be generated by the jammer to provide adequate power density over the radar band.

DECEPTION JAMMERS

Deception jammers are far more conservative in terms of power required to accomplish their mission. Rather than "out-shout" the radar return as in the case of noise jammers it is the function of deception jammers to confuse the operator by providing "false" information. For a radar to direct a fire-control system correctly it must locate the target and then establish range and bearing to the target. Hence, if either range or bearing, or both, are misrepresented without the operator's knowledge the target's location will be established incorrectly.

The radar determines the range of a target by measuring the length of time it takes for a pulse of energy to travel from the radar antenna to the target and return. This time interval is measured with reference to a local signal that opens a "range gate" during the time interval in which the return signal is received. As the target moves closer, or farther away, the opening of the range gate is advanced, or retarded, in time. It is now possible for a jammer to provide a false-echo return slightly greater in amplitude than the true echo. The time period of transmission for this false echo is then varied so it will appear that the target is moving in some other direction than on its true course. The radar range gate tends to shift, having locked onto the stronger false signal. This shift, of course, presents the incorrect target range to the radar operator; the jammer capable of doing so is called a *range-gate stealer*.

A second method of confusing the radar operator is to provide incorrect target-bearing information. In its simplest form this method amounts to providing a false signal to the radar when the antenna is pointing in a direction other than at the true target. To accomplish this confusion successfully, it is necessary for the jammer system to sense the scan pattern of the radar. Once the pattern is analyzed, it is possible to know when the radar antenna is pointed away from the true target. The jammer then sends out a strong false echo signal, which the radar operator now interprets as being at the bearing on which he is presently looking. One typical jammer system that basically uses this technique is known as an *inverse conical scan repeater* and will be discussed in detail in Chapter 6.

The last deception jammer to be considered here is the *false-target generator*. This technique generates many false echoes, which vary about the real target echo in both range and azimuth. To accomplish this effect, the jammer system requires additional complex circuitry, but less average power than noise jammers. As in the previous systems it is necessary that the jammer system establish the radar search pattern and use this information to generate the jamming signals.

It can be seen from the foregoing discussion that deception jammers use far more sophisticated circuitry and techniques than noise jammers do. Where noise jamming tends to use a brute-force blanket effect, deception jammers tend to deprive the radar of information associated with one of the parameters (range or bearing) necessary to locate the target in space.

OTHER RADAR CONFUSION DEVICES

There are at least three other major approaches to offsetting the radar-detection capability, which we mention briefly here. The first and perhaps the oldest ECM technique is *chaff*. Chaff consists of pieces of tinfoil or other lightweight reflecting material that can be dispensed by an aircraft in flight. The length of the chaff, which is generally in ribbon-like pieces, is selected to give a good radar return over as large a portion of the radar band as possible. Often the length will simulate a half wave at, or close to, the radar frequency of interest. This chaff is dispensed in large quantities when the aircraft is being tracked by unfriendly radar. Because of its light weight it tends to remain in the air behind the aircraft and float slowly to the earth. The radar echo returned from the chaff is generally stronger than the echo returned from the aircraft and therefore the radar begins tracking the chaff and "loses" the target aircraft. Of course, the chaff tends to lose its forward velocity after a few moments and is also falling in space. These inconsistencies as compared to aircraft flight shortly become apparent to the radar operator and a new search for the real target will be initiated. However, the time gained by the aircraft may be very useful.

Another way in which chaff is used is called "chaff-corridor seeding." In this method the lead aircraft, flying at a slightly higher altitude than the following planes, dispenses large quantities of chaff in a corridor fashion as it approaches the radar. As the chaff floats down, the following aircraft are provided with a screen that masks their approach. The search radar will receive one large return from this quantity of chaff. It may either discount this return as simply the result of the first aircraft's ECCM tactics, or at best may be able to define planes in the chaff corridor, but generally will not be able to establish the exact number.

The *decoy* is not generally thought to be in the same class as chaff. Nevertheless, the two have many characteristics in common. The decoy may be passive or active in nature. In either case, its primary function is to give the impression of an approaching aircraft to the radar display. Since the decoy is provided with primary power of its own, its velocity can be at, or close to, that of the parent aircraft. It is impossible to distinguish between the true target and the decoy on the basis of velocity differential. Although each aircraft, because of weight limitations, can carry only one or two decoys it is generally conceded that each decoy will draw a defense weapon to its destruction. If to every decoy carried by the incoming raid an interceptor must be assigned, or a ground-to-air missile expended to destroy it, then the added weight is well worth the price.

There is one additional approach to providing protection from a searching radar. The methods previously discussed admit to the existence of a hostile target but attempt to deprive the defense fire-control system of the information necessary to effect a kill. *Radar-absorbing materials* make it possible to hide the incoming aircraft completely by reducing the radar echo to a very weak level. These radar-absorbing materials have the same effect on the radar return as sound-absorbing ceilings have in soundproofing a room. With this approach it is hoped that the defense will never be aware of the presence of an enemy aircraft.

It can be seen from this discussion that there are three general classes of ECM devices. These are active noise jammers, active deception jammers, and passive devices. The first requires considerable primary power and employs a brute-force technique. The second is somewhat more sophisticated, depending on complex circuitry rather than brute power to provide effective jamming. The last group requires no primary power but consumes a large portion of the aircraft's net payload. As is generally true, all of one's eggs are not put into a single basket, but rather a mix of these various techniques is employed in a given raid. The factors affecting this mix depend on the defense system to be penetrated. A detailed discussion of a logical approach to the selection of the ECM equipment mix is taken up in Chapter 7.

There is one additional mode of operation for a fire-control system that is of special interest to the ECM field. This technique is known as *home-on jam*. In this mode the fire-control radar does not depend on its own return for a signal but instead locks onto the jamming signal. This technique is particularly effective against noise jamming. Hence, when the noise source destroys the normal radar echo the fire-control radar switches over to the "home-on jam" mode and directs its missile or interceptor directly at the jammer source. This approach can make jamming a very hazardous technique. However, the jammer may still exercise at least one of two protective plans. First the jamming aircraft may also monitor the area by using airborne radar. This method allows the observation of the approach of a "home-on jam" missile. With this information the jammer may be turned off at a critical time in the missile-flight plan, thus destroying the guiding source.

The second approach is to provide a *stand-off jamming* source. In this case the jamming aircraft remains outside the maximum range of the defense weapon while it provides a strong, continuous jamming signal. Weapons capable of the home-on jam mode of operation are now attracted toward this source. However, since this aircraft is not approaching the target area as expected, the defense weapon will expend its fuel, or must turn back, before engaging the enemy. Obviously, while this action is going on, "quiet" aircraft, with no jamming, are attempting to slip through the defense system.

TYPICAL AIR-COMBAT ANALYSIS IN AN ECM ENVIRONMENT

Assuming that the ground radars can locate and track the incoming bombers, the interceptor can then be directed to the raid area as discussed earlier. As already stated, it would be impractical to treat here all possible modes that can develop. For that reason, only a co-altitude pursuit mode of attack will be considered and a few of the elementary analytical steps will be carried through to illustrate the major parameters that must be considered.

The geometry of the problem is shown in Fig. 2–1. The target (bomber) is at the origin of the polar plot. The interceptor is shown approaching directly from a tail-chase position. Speaking more generally, it should be understood that the interceptor could approach its target from any azimuth. However, for attack modes other than pursuit the relative paths between the two planes are continually changing. The interceptor computer must, therefore, receive continual information from its fire-control radar to correct its course. In a jamming environment this information is at a premium. From this point of view, the pursuit mode, which requires less information for its successful execution, is more desirable.

In Fig. 2–1 two "barrier" curves are also shown. The curve labeled "launch barrier" defines the last possible chance the interceptor pilot has to launch his air-to-air missile. For the case shown here it is about 3000 ft from the target. If he approaches closer to the target before firing, his weapon may not have time to respond properly or he may actually crash into the bomber himself.

The "radar-barrier" curve defines the minimum distance allowable for acquiring his target on the radar screen. The difference between the two barriers provides for the pilot reaction time (in the order of 2 to 3 sec) and weapon selection and preparation time. The radar barrier in this case is shown to be at about 8000 ft along the interceptor line of attack.

The target size and the angle of approach provide a certain reflective area for the radar signal. For the situation to be considered here this area will be assumed to equal 10 sq m. In practice this area may vary anywhere from 1 to over 100 sq m as a function of this approach angle and target size.

In a "clean" (non-jamming) environment 10 sq m provides a radar

18 TECHNIQUE AND TACTICS

reflection quite adequate for detection of the target at a distance in excess of that needed (8000 ft in this case) by the interceptor. The point of interest here is to see what effect jamming has on this requirement. Bear in mind that the interest is in understanding the general steps in the problem rather than a detailed discussion of the radar range equation at this time. This radar-range problem will be discussed in considerable detail in later chapters.

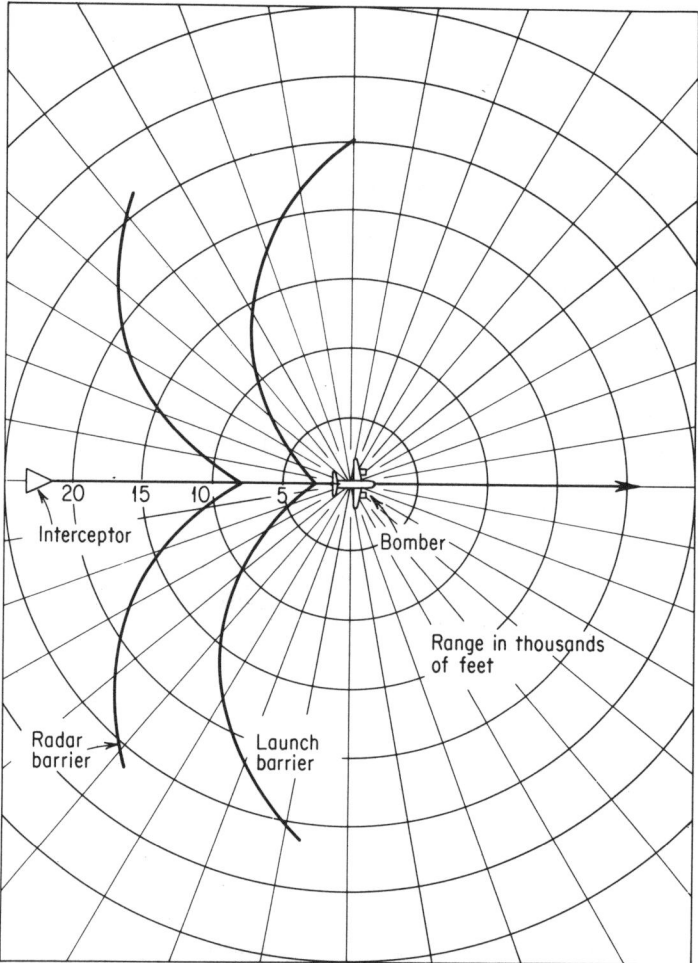

Figure 2-1. *Attack geometry for pursuit mode discussed in text.*

To make the problem clear it will be assumed that the bomber carries a noise jammer capable of generating a power density of 1.0 w/mcs over the frequency range employed by the interceptor radar. This would be the desired result of a typical tactic employing a barrage-noise jammer as discussed earlier. It is obvious that the radar must enjoy a certain radar-signal strength to noise- "signal" strength ratio in order to detect its target. This relationship is exactly analogous to the station-signal strength on a home radio competing with "static." When a bad lightning storm generates a great deal of noise the desired signal cannot be heard (detected). This ratio can be expressed for the radar/jammer case as:

$$\frac{S_r}{S_j} = \frac{\text{radar signal}}{\text{jammer noise signal}}.$$

When dealing with typical pulse radars this ratio applies to the return from a single pulse and will be referred to as

$$\left[\frac{S_r}{S_J}\right]_{\text{Effective}}$$

Fortunately for the interceptor, this ratio is enhanced for the model being considered by a process known as "pulse integration."* The interceptor radar transmits a series of pulses as it approaches its target and the pilot attempts to establish the bomber's position by the location of a returned pulse on the radar screen. If he had to depend on seeing just one return "blip," as it is called, it is quite possible he would miss it. However, a number of such blips are returned and his eye tends to integrate the result, providing a more positive target location. This improvement, or the number of blips that can be expected for identification, is a function of the closing velocity, V_c, between the interceptor and the bomber. To understand this effect, assume for the moment that the interceptor closed on the bomber with an infinite speed. Since radar waves propagate at a finite speed (186,000 miles/sec), there would not even be time for one pulse to be returned. On the other hand, if the interceptor just followed the bomber at a fixed distance for an extremely long period there would be sufficient time for an infinite number of pulses to be reflected. Since the interceptor travels faster than the bomber the closing rate can be expressed as $V_c = V_I - V_B$, for the simple pursuit mode being considered here. If the interceptor velocity, V_I, is 900 mph (1300 fps) and the bomber velocity, V_B, is 700 mph (1000 fps), the closing rate between the two is 300 fps in a tail-chase condition, and 2300 fps for a head-on condition. This effective number of pulses for integration is then given by:

$$N_{EF} = \frac{0.85\,\tau c}{4V_c} \cdot \text{prf} \qquad (2\text{--}1)$$

where τ is the radar pulse length in seconds,

c is the velocity of propagation of electromagnetic energy,

prf is the pulse repetition frequency of the radar,

V_c is the closing rate between interceptor and target.

* W. M. Hall, "Predications of Pulse Radar Performance," *Proceedings of the IRE*, Feb. 1956.

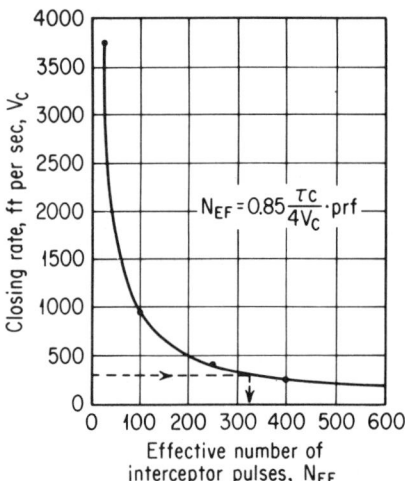

Figure 2-2. *Effective number of pulses for integration, N_{EF}, as a function of closing rate, V_c, between target and interceptor.*

This relationship is plotted in Fig. 2–2 for a typical radar with a prf of 900 pulses/sec. For a closing rate of 300 fps between the two aircraft, the effective number of pulses, N_{EF}, available for integration is shown to be 325. To establish the signal enhancement to be realized by this "integration" it is first necessary to determine the initial signal-to-jam ratio that exists at the point of interest. This ratio is determined in Fig. 2-3, which shows a plot of $[S_r/S_j]_{EF}$ existing at various distances from the target as a function of jamming-power densities. Since the minimum range at which radar crossover must be acquired is 1.32 nautical miles and a jamming density of 1 w/mcs is specified the effective S_r/S_j is seen to be −2 db. The pulse integration enhancement that can be expected for this condition is given by:

$$\left[\frac{S_r}{S_j}\right]_{IF} = (N_{EF})^{0.8} \left[\frac{S_r}{S_j}\right]_{EF} \qquad (2-2)$$

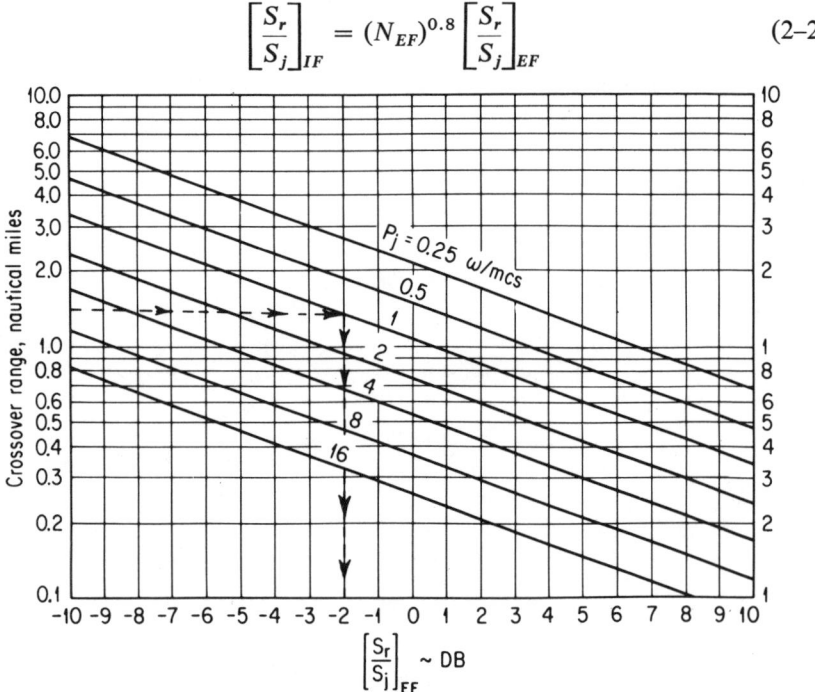

Figure 2-3. *Crossover range vs. $[S_r/S_j]_{EF}$.*

This equation is plotted in Fig. 2–4 for various levels of integrated pulses. For 325 pulses the signal improvement ratio, $[S_r/S_j]_{IF}$, is seen to be about 16 db.

Finally, from Hall's* plot of the probability of detection as a function of

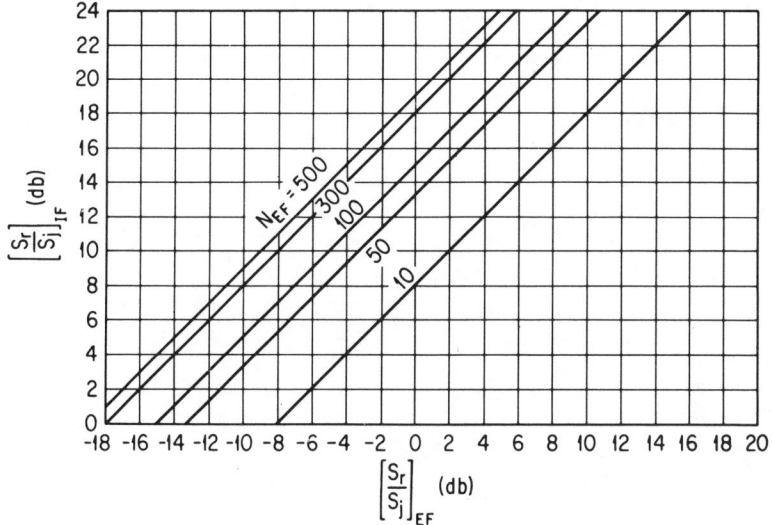

Figure 2-4. *Signal-noise ratio improvement due to integration as a function of N_{EF}.*

the signal-to-noise ratio, the desired probability of the interceptor detecting its target at the critical distance of 1.32 nautical miles is 0.999 (Fig. 2–5).

To fully appreciate the effect of jamming on the probability of the interceptor's detecting its target, consider the same situation with an increased jamming density of 4 w/mcs. The $[S_r/S_j]_{EF}$ at 1.32 nautical miles is now -8 db, the $[S_r/S_j]_{IF}$ is 10 db, and the probability of detection is reduced to 0.25. This simplified example serves to show the analytical technique used in establishing the effectiveness of a jamming environment on the terminal phase of an interceptor attack.

In this section we have given a quantitative example for the evaluation of a specific technique, namely barrage noise jamming. The next section will provide an analytical example for the evaluation of a specific tactical problem.

TACTICS—AN EXAMPLE

In employing ECM, adequate consideration must be given to the advantages as well as the disadvantages it affords the enemy. These considerations in their proper perspective give rise to the choice of tactics. The advantage given to the enemy arises from the fact that a system must perform several functions in order to accomplish its mission. ECM may, while confusing one function,

* See reference on page 19.

22 TECHNIQUE AND TACTICS

actually expedite the accomplishment of another. For example, active ECM may degrade the tracking function of a defense system but at the same time it insures detection by the defense system and eliminates indecision as to the identity of the user as hostile. If the ECM is activated too early it increases

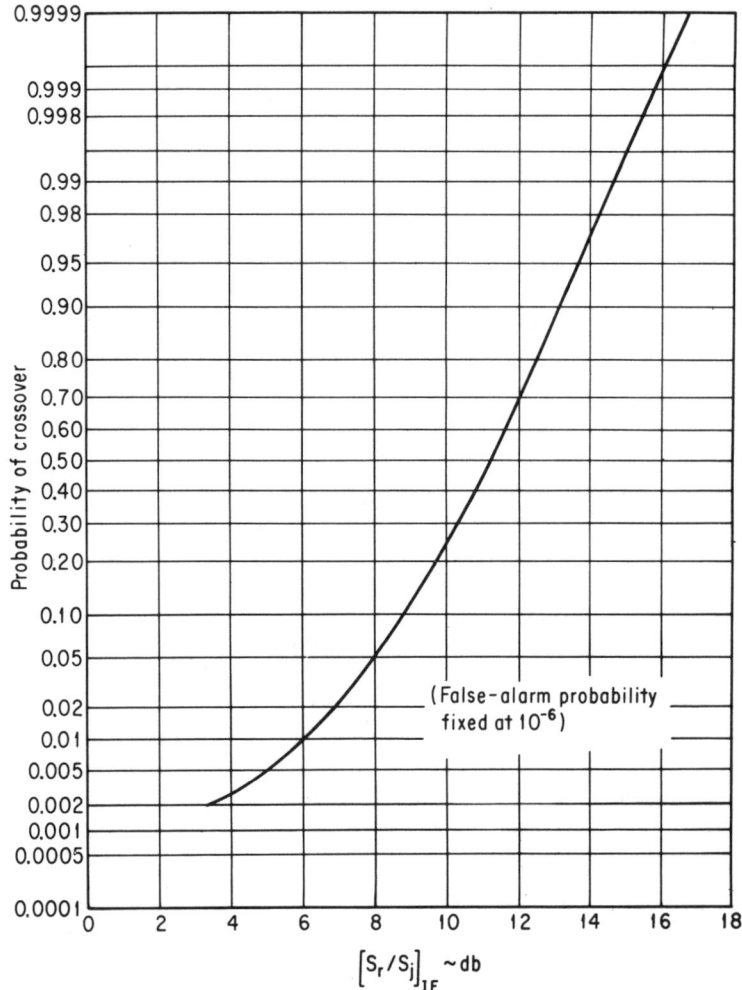

Figure 2-5. *Probability of crossover at range R_{CO} vs. $[S_r/S_j]_{IF}$.*

the range at which the defense system can launch an attack. If it is used too late in the attack, the defense system may not be degraded in its tracking capability until it has successfully launched an attack. Thus, there exists an optimum time at which to activate the ECM and the value of this time and the sensitivity of defense effectiveness to the appropriate choice of time depends critically on the nature of the defense system.

TACTICS—AN EXAMPLE

To illustrate the choice of timing, consider a target defended by a surface-to-air missile (SAM) system. Assume that the system has the capability of determining whether or not a raid is within missile range (triangulation) and has a home-on-jam mode of operation. The home-on-jam mode will generally have a lower probability of destroying a member of the raid than the normal operating mode.

The effectiveness of the defense system is reflected by the number of salvoes it can launch against the raid, times the kill probability of a missile in the salvo.* Let this product be E, then

$$E = N(R) \cdot P(R)$$

where $N(R)$ = number of salvoes if the ECM is activated at range R from the target,
$P(R)$ = average kill probability per missile launched given the range of activation R.

In general, $P(R)$ will vary from salvo to salvo, especially if the ECM is activated after the attack by the defense is launched.

Now, it can be shown that the number of salvoes that can be launched is:

$$n = 0; \quad R_1 < R_b$$

$$n = 1 + \left[\left\langle \frac{\ln\left(\frac{R_1 + ut_1}{R_b + ut_1}\right)}{\ln(1 + (v/u))} \right\rangle \right] \tag{2-3}$$

where R_1 = range of interception of first salvo,
R_b = range to bomb-release line (i.e., the maximum distance that the raid may be from a target to achieve a hit),
u = SAM speed,
t_1 = time delay from when a salvo meets the raid to launching of the next salvo,
v = raid speed,
$\langle x \rangle$ = integral part of x, i.e. $\begin{vmatrix} \langle 1.4 \rangle = 1 \\ \langle 0.7 \rangle = 0 \end{vmatrix}$.

The range of interceptor of the first salvo, R_1, is given by

$$R_1 = \min\left[R_m, \frac{R_d - vt}{1 + (v/u)} \right] \tag{2-4}$$

where R_m = maximum SAM range,
R_d = range of detection,
t = time delay between detection and launch.

---

* Note that the problem is much more complicated in reality, since it depends on the ratio of missiles to raid members. However, for illustration purposes this simplification is adequate.

24 TECHNIQUE AND TACTICS

Now if ECM were not used, detection would occur at a range $\bar{R}_d$, if it is employed at a range $R > \bar{R}_d$, then detection will generally occur at R. Thus

$$R_d = \max [R, \bar{R}_d]. \qquad (2-5)$$

Equations 2–3 through 2–5 enable one to determine the number of salvoes as a function of the range of activation of the ECM. It should be obvious that the optimum range of employment of the ECM is $\bar{R}_d$, since at this value the ECM does not increase the number of salvoes but does act to decrease the probability of a kill of all salvoes fired. However, since detection is a stochas-

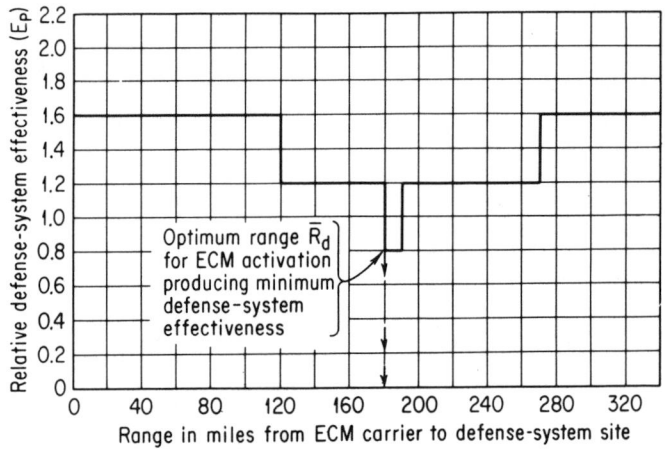

Figure 2-6. *Relative defense-system effectiveness, E_R, vs. range of ECM activation.*

tic process and since $\bar{R}_d$ would not be known exactly to the raid even if it were deterministic, the value of R will be chosen with some error. Figure 2–6 illustrates the effect of the choice of R for the set of defense-system and raid characteristics shown in Table 2–1.

Table 2–1. *Assumed defense-system and raid characteristics*

Term	Symbol	Value
Maximum SAM range	R_m	180
Detection range without ECM	R_D	180
Bomb release range	R_b	50
Initial time delay (detection to launch)	t	1
Time between salvo intercept and launch	t_1	0.5
SAM speed	u	50
Raid speed	v	20
Probability of kill without ECM	p	0.8
Probability of kill with ECM	p	0.4

The figure illustrates a great sensitivity to the choice of R, and this sensitivity depends on the many factors discussed. If, for example, the defense system had no home-on-jam capability but had to "burn through" the ECM, the effectiveness would be constant at some low level as long as R were greater than $\bar{R}_d$ but would increase as R fell below $\bar{R}_d$. If the defense system could not determine the range at which firing could be made, the ECM could cause an exhaustion of the missile supply or cause the defense to hold fire for too long a period. The major factors are that choice of tactics can be extremely important and that the proper choice is integral to the technique employed and to the defense-system and raid characteristics.

3

NOISE, PROBABILITY, AND INFORMATION RECOVERY

RELATIONSHIP OF "NOISE" TO AN INFORMATION-CARRYING SYSTEM

It is desirable to preface a detailed examination of the problem of electronic warfare with a brief consideration of several physical concepts fundamental to the issue. First, the general subject of noise in an information-carrying system will be discussed. Noise exists to a greater or lesser extent in all electronic systems; it is therefore logical to consider the general problem first, and to treat the intentional introduction of noise (i.e., as by a hostile noise-jamming transmitter against radio or radar) as simply an extension of the general case. It will be seen that the presence of noise in a system introduces uncertainty as to the information content and thereby degrades system performance as measured in terms of accuracy, reliability, information-handling capacity, or some similar criterion. It is therefore obvious that the intent (and the effect) of jamming is to deprive the victim of the use of the full capabilities of his electronic system(s). In order to consider in some detail the relationship between noise and information recovery, it is first necessary to introduce some elementary probability theory and associated statistical concepts. This discussion is provided in the present chapter to support the remainder of the text material. More detailed treatment of the problem of retrieving a signal from an ECM (noise) environment is presented in Chapters 4, 5, and 7.

ELEMENTARY TREATMENT OF NOISE THEORY

The so-called "noiseless channel" is seldom, if ever, encountered in physical reality. It is one of life's hard facts that where an information-bearing signal is found, it is always associated to some extent with an environment of competing "signals." In many cases, as, for example, the familiar occurrence of atmospheric "static" interfering with radio reception, the environment from which the desired information must be extracted is of an irregular nature popularly referred to as "noise." On other occasions the desired signal may be required to compete with one or more information-bearing, but undesired, signals of a nature similar to its own. For purposes of semantic clarity, it is conventional in the discussion of information theory to refer to an emanation whose reception is desired for the purpose of information extraction as a "signal," regardless of its physical composition. Conversely, any other emanations with which the "signal" must compete for recognition or reception are classified as "noise." Thus, whereas electromagnetic radiations from a celestial body may represent "noise" on the desired target echo "signal" received at a modern radar, the radio astronomer might well be plagued by "noise" from a near-by radar as he attempts analysis of weak stellar "signals" from a distant galaxy.

The fundamental problem of information recovery from a signal immersed in noise is essentially independent of the exact nature of either signal or noise, at least within the framework of the considerations of "electronic warfare." Among point-to-point communications systems, the simplest are those utilizing some form of "binary" information transmission, i.e., those which transmit a signal alternating between two possible states. Radio-telegraph systems and some simple digital data links fall into this category. Each may use, for example, a Continuous-Wave (*CW*) radio-frequency carrier that, at any instant, must be either "on" or "off." The use of such an interrupted carrier is merely an extension of the use of interrupted, or keyed, d-c voltage on a wire circuit. Alternatively, the *RF* carrier may be shifted between two discrete frequencies, as in Frequency-Shift-Keying (*FSK*) commonly used in radioteletype work. In each case, the problem of information recovery is basically that of determining, at certain intervals in time, the presence or absence of a specific signal or signals. The signal must be distinguished from undesired "noise" by recognition of characteristic signal parameters, such as frequency, time of occurrence, duration, amplitude, and so on. Remote-control applications, such as are encountered in aircraft or missiles command-guided from the ground or from other aircraft, are variations of point-to-point communications links, and the problems of recovering information-bearing signals from a noise environment are identical.

Simple pulse-radar systems transmit an interrupted *RF* carrier, and the presence or absence at any instant of a reflected echo in the receiver, usually but not necessarily located at the transmitting site, provides information relative to the existence and position of reflective targets, if any, within range of the radar. Despite differences in geometry of the respective propagation paths, the simple radar system is thus identical from the information-recovery standpoint with the point-to-point communications link. More complex radars extract additional target information by analysis of modulations imposed on the signal in the process of reflection by the target. The recovery of such information may be considered as imposing an additional restriction or group of restrictions on the definition of a "signal." If, for example, it is desired to determine the radial velocity of an aircraft target in relation to the radar site, the information-recovery problem may be phrased as follows: for a series of Doppler-frequency shifts f_i (change in received carrier frequency due to reflection from a radially moving target) corresponding to a series of radial velocities v_i, does a received signal with Doppler shift f_i exist at any given instant? Thus, for each receiver channel corresponding to a given received RF frequency (or range of frequencies), the problem reduces to the familiar one of determining the presence or absence of a carefully defined signal at any given instant. Other problems in the analysis of unique signal characteristics may be similarly reduced to the simple basic form, "Is or is not a signal possessing specified characteristics present at time T_1? T_2?..., etc."

Even the detection of voice modulation in noise may be reduced to these same terms. Consider the audio signal to be transmitted and subsequently received as an amplitude function, A_i of successive discrete time intervals, T_i, such that $A_i = a(T_i)$. That is, an approximation to the continuous (audio) function of time is made as a series of step functions, with each amplitude A_i corresponding to the time interval T_i. Let the instantaneous RF signal parameter (amplitude, frequency, and so on) corresponding to each A_i be $X_i = x(A_i)$. Further, divide the range of possible amplitude values of A into equal discrete intervals, designated A^j, with corresponding intervals of the relevant modulation parameter designated X^j. Thus, X_9^7, for example, would designate that the value of the modulation parameter during time interval No. 9 lay in the seventh range, which might correspond to an instantaneous amplitude between 70% and 75% of peak amplitude. Now it may be asked, "What signals X_i^j are present in the received wave?" Phrased in more obvious language, "During each time interval, T_i, which (if any) of the possible values of X^j are present?" Clearly, this is essentially the same question asked earlier in the case of a simple binary system, although the complexity of equipment required to instrument the voice-detection channel in this form is much greater than for the binary systems. It is also apparent that the specification of discrete time and amplitude values above, though a workable approach in fact, might, if desired, be extended to provide in the receiver an exact duplicate of the original audio wave (provided all decisions as to the presence of X_i^j are made correctly). It is only necessary to allow the time intervals between T_i's, and the functionally related intervals between A^j's and corresponding X^j's, to decrease toward zero in the limit.

It has been shown, somewhat intuitively, that in any information-transmission system the problem of recovering the desired information from the received signal when the latter is immersed in a competing noise environment may be reduced to one fundamental question: "For each possible combination of information-bearing signal parameters, is or is not such a signal present?" Since the "at any given time" added in previous statements of the above question is in reality one of the possible information-bearing parameters of the signal, it may be omitted from explicit mention. Indeed, in some applications the time of signal arrival does not yield significant information.

At this point it will be profitable to restrict the discussion to cases and problems of interest within the framework of "electronic warfare." We thus rule out consideration of wire systems, which are intrinsically impervious to the techniques of electronic warfare because of the difficulty of opposing forces injecting signals or noise into each other's wire links. In order to preclude the necessity for speaking in uncomfortably general terms, the comments will be confined to a discussion of information recovery with respect to relatively simple pulse radars. Not only are a substantial proportion of practical problems actually related to such radars (as encountered, for example, in air defense and aircraft/missile-guidance systems), but also, as preceding paragraphs have shown, detection problems in most other systems may be considered as extensions of the simple binary case typified by an elementary pulse radar.

Given a theoretical, noiseless information-transmission channel, the process of recovering the desired information from the received wave would indeed be straightforward. Since, by definition, the desired signal and only the desired signal could exist at the channel output, it might be amplified (in noiseless amplifiers, of course) to the extent necessary for operation of whatever device might be chosen to display or process the output. In any practical channel, however, a number of noise sources enter the picture. Noise is generated in every electronic circuit, be it only a few inches of wire, by thermal motion of free electrons in the conductors. In addition to this thermal noise, inherent even in passive circuits, every device that depends upon the motion of charged particles, e.g., electrons, is a noise source. Thus conventional vacuum tubes generate thermal noise due to irregular thermal velocities of electrons emitted from the cathode, shot noise due to arrival of individual electrons at the anode, and flicker noise due to uneven and fluctuating electron emission from different areas of the cathode. Solid-state devices, klystrons, traveling-wave tubes, and in fact every related variety of active electronic device, are also sources of electronic noise. Last, but of particular interest in the discussion of electronic warfare, is noise originating outside of the transmitter and receiver proper, but entering the receiver via the antenna. In the latter category are natural noise radiations of galactic or atmospheric origin, and man-made noise, either accidental or intentional.

In the unlikely event that there were available, at the receiving end of an information-transmission circuit, full knowledge of the detailed characteristics of all interfering noise (particularly of intentional noise-jamming sources), it would then be possible, at least in principle, to detect the desired message without error. The noise might simply be subtracted from the known composite noise waveform of the total received wave, leaving only the original transmitted signal. Indeed, the noise-cancellation technique described at the end of this chapter operates on this same principle to largely eliminate some types of high-amplitude interference originating outside the transmitter-receiver equipment (e.g., atmospheric "static").

In the majority of cases, it is not practical to implement this cancellation process physically. With the exception of some man-made types, all the noise varieties mentioned earlier are the product of *stochastic*, or chance, processes. This origin is perhaps best illustrated by thermal noise resulting from the random variable of electron motion.

A *random variable* is one resulting from or originating in a stochastic process, and may be defined as one *whose instantaneous values are independent (not related in any way) from any instant to any other*. Random noise, then, is noise originating in a random process, and its amplitude at any instant is, by definition, unrelated to its amplitude at any earlier or later time. This characteristic is well illustrated by thermal noise, resulting from the thermally excited motion of free electrons. The free electrons within a conductor possess individual velocities (at any instant), which are the result of astronomical numbers of chance collisions (a stochastic process) among themselves. Thus both the distance (free path) an electron travels between collisions and its velocity are random variables. It cannot, therefore, be forecasted in detail what the future of individual electrons will be. However, the *average* distribution of velocities among a large number of these particles follows a very definite pattern, as does the average distribution of free-path lengths. That is, it is possible to describe the average behavior of the electrons on the basis of the macroscopic characteristics of the process. Although it is not possible to speak cogently of the behavior of individual electrons emitted from a thermionic cathode, it is possible to make a number of statements on a statistical basis. The average velocities and free-path lengths of free electrons within the cathode are functions of absolute cathode temperature. Given a sufficient number of emitted electrons, m, it is possible to establish that at some temperature, T, n electrons, on the average, will be emitted with velocities between v_i and $v_i + \Delta v$. In the language of statistics and probability, this statement may be rephrased as follows: At temperature T, the *probability* that an emitted electron will possess velocity between v_i and $v_i + \Delta v$ is n/m. Thus, on the basis of the past observed behavior of certain statistical properties of a large number of electrons emitted from a thermionic cathode, confident prediction of the future *average* behavior of these same properties can be specified.

ELEMENTARY PROBABILITY AND STATISTICS

It is now appropriate to introduce certain elementary relationships in probability theory. These will be applied on occasion throughout the text and extended somewhat where necessary. This material is introduced here to familiarize the reader with some of the mathematical background required in order to fully comprehend the topics to be treated and to provide some preparedness for these problems. Obviously, all of the theory of probability cannot be, nor need it be, covered here. It is also intended that sufficient treatment will be provided to acquaint the non-mathematical reader with the terminology involved. This background will allow an understanding of the text material in later chapters, where derivations may be omitted if desired.

Suppose that a quantity of balls are numbered consecutively, 1 through m, and are placed in a container and thoroughly "scrambled." The container is just large enough to contain all m balls in a single layer, and has m slight depressions, also numbered 1 through m, into which the balls settle after scrambling. The probability of one ball falling in any given depression is *one*, or unity, and similarly the probability of a ball being in every one of the m depressions is also one, since the probability of occurrence of any event that is certain to take place is, by definition, one. Conversely, the probability of any depression remaining empty is zero, since, by definition, the probability of occurrence of an event that cannot take place is zero. More generally, the probability of any particular outcome of a specified act is equal to the number of times, on the average, that that outcome will occur, divided by the total number of repetitions of the act. A more mathematically useful statement is that the probability of an event's occurrence is equal to the number of different ways in which the event may possibly occur, divided by the total number of equally likely outcomes of the situation. Any one of the m balls in the above example is equally likely to fall in any one of the m depressions. There is obviously only one way in which a specified ball may end up in any given position, hence the probability of a given ball falling in a given spot is $1/m$. If the depressions are arranged in rows of n depressions each, then the probability of any given ball falling in any given row is n/m.

In this example, the final positions of the m balls are not independent outcomes since they do not result from independent events. That is, if ball No. 1 falls in position No. 4, then there are only $m - 1$ remaining positions into which the remaining balls can fall, hence the probability of any other ball falling in position No. 2 is $1/(m - 1)$, and so forth. It may be stated that the joint probability of two non-independent outcomes, A and B, is equal to the probability of A times the conditional probability of B given that A has occurred. Symbolically,

$$P(A, B) = P(A)P_A(B). \tag{3-1}$$

If the outcomes are independent, i.e., if the occurrence of A does not affect the probability of occurrence of B, then $P_A(B)$ becomes simply $P(B)$; the joint probability of independent events A, B, ... r, is

$$P(A, B, ..., r) = P(A)\, P(B) \, ... \, P(r) = \prod_{i=A}^{R} P(i). \qquad (3\text{-}2)$$

If there exist several *mutually exclusive* possible outcomes of an act, then the cumulative probability that *either* outcome A or outcome B will occur is

$$P(A \text{ or } B) = P(A) + P(B). \qquad (3\text{-}3)$$

If two conventional six-sided dice are thrown, there are 6 equally likely outcomes for the first die (1 through 6), and, independently, 6 for the second. There are thus 36 possible permutations of the two dice. There are, however, only 11 *different combinations* possible, since the final outcome must be an integer, x, between 2 and 12; any outcome between 3 and 11 may result from two or more permutations, as tabulated in Fig. 3–1.

x	$P(x)$	n	Permutations yielding x
0	0	0	None
2	$\frac{1}{36}$	1	1-1
3	$\frac{2}{36}$	2	1-2 2-1
4	$\frac{3}{36}$	3	1-3 2-2 3-1
5	$\frac{4}{36}$	4	1-4 2-3 3-2 4-1
6	$\frac{5}{36}$	5	1-5 2-4 3-3 4-2 5-1
7	$\frac{6}{36}$	6	1-6 2-5 3-4 4-3 5-2 6-1
8	$\frac{5}{36}$	5	2-6 3-5 4-4 5-3 6-2
9	$\frac{4}{36}$	4	3-6 4-5 5-4 6-3
10	$\frac{3}{36}$	3	4-6 5-5 6-4
11	$\frac{2}{36}$	2	5-6 6-5
12	$\frac{1}{36}$	1	6-6
13 up	0	0	None
Total	$\frac{36}{36}$	36	

Figure 3–1

For each possible outcome, x, of the dice, the column, n, indicates the total number of permutations of the two that may yield the given x. The permutations are tabulated at the right. The total number of possible permutations, which is the sum of column n, is $6^2 = 36$. Thus, the probability of any particular outcome, x, is $P(x) = n/36$, as shown.

Since the 11 possible outcomes, 2 through 12, are mutually exclusive, the cumulative probability that one of them will occur is the sum of their individual probabilities. From Eq. 3–3,

$$P(2 \text{ or } 3 \text{ or } ... \text{ or } 12) = P(2) + P(3) + ... + P(12) = 1. \qquad (3\text{-}4)$$

This result is expected since some outcome between 2 and 12 is bound to occur, hence the cumulative probability must be unity. The discrete probability distribution of $P(x)$ is given graphically in Fig. 3–2.

It is now desired to apply the probability-distribution theory to physical random-noise problems. Continuing with the example of free electrons emitted from a thermionic cathode, it is known that these particles may possess linear velocities of any value within a very wide range. They are not limited to any finite number of possible discrete velocities, hence the velocity probability distribution function, $P(v)$, is continuous, i.e., v is said to be a *continuous variate*. Since v may attain any of an infinite number of values between any two discrete limits, and since the sum of $P(v)$ for all possible v must equal unity (an electron must of course possess *some* value of v), it is apparent that a modified definition of $P(v)$ is required. Therefore, by definition the probability of v lying in the incremental region between v_i and $v_i + \Delta v_i$ is $P(v_i) \cdot \Delta v$. Furthermore, if the entire possible range of v is divided into intervals of width Δv beginning at $v_1, v_2 \ldots v_s$, then

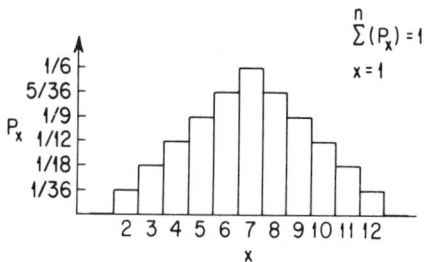

Figure 3-2. *Discrete probability distribution, P_x, for the throw of two six-sided dice.*

$$\sum_{i=1}^{s} P(v_i) \Delta v = 1. \qquad (3-5)$$

By allowing Δv to go toward zero in the limit, it is possible to replace summation over v_i by integration over all v, thus:

$$\int_{-\infty}^{+\infty} P(v) \, dv = 1. \qquad (3-6)$$

$P(v)$ as defined here is now a continuous function of v, such that the probability of a single electron having a linear velocity between v_1 and v_2 is

$$P(v_1, v_2) = \int_{v_1}^{v_2} P(v) \, dv. \qquad (3-7)$$

The discrete probabilities discussed previously in connection with balls and dice were arrived at by deductive reasoning, and as such are known as *a priori* probabilities. In the case of free electrons, there is no such obvious reasoning process by which it is possible to arrive at a "logical" velocity-distribution function. It has been experimentally determined, however, that the actual distribution is of the form

$$P(v) = K_1 e^{-K_2(v-\mu)^2}. \qquad (3-8)$$

This is the so-called normal or Gaussian distribution function, and is shown plotted for one pair of values, K_1 and K_2, in Fig. 3–3. The Gaussian distribution is symmetrical about its *mean* value, μ, and describes the distribution of continuous variates in a remarkable number of problems involving stochastic (random) processes. Its exact shape depends, of course on the constants K_1 and K_2, as well as upon μ. It is often convenient to write the normalized Gaussian distribution function,

$$P(v) = \frac{1}{\sqrt{2\pi}\sigma} e^{-(v-\mu)^2/2\sigma^2}. \qquad (3-9)$$

The parameter, σ, is the standard deviation of the distribution, and determines its relative width and height. σ^2 is the mean squared deviation of the distribution about the mean, μ, and is useful as a measure of noise power or energy in many applications. $P(v)$ does not go to zero for any finite value of v, indicating in the present example the finite probability of finding a small number of electrons with very high "positive" velocities (i.e., toward the anode), as well as some with large negative velocities (toward the cathode). It can readily be shown that Eq. 3–9 above satisfies the requirement of Eq. 3–6 for all values of μ and σ. The mean velocity, μ, corresponds in the case of a vacuum diode to an average (d-c) current flow.

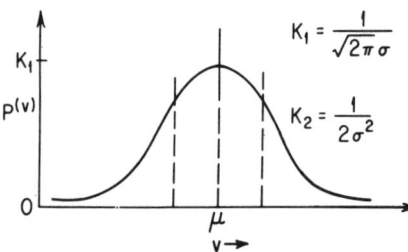

Figure 3-3. *The normal (Gaussian) probability distribution function,*
$$P(v) = K_{1e}{}^{-K_2(v-\mu)^2}.$$

Several important characteristics of probability distribution functions relate to the addition of two or more such independent distributions, as in the addition of signal and noise or of two statistically independent random noise signals; the mean value and the mean squared deviation of the new, resultant probability distribution are equal, respectively, to the sum of the mean values and the sum of the mean squared deviations of the constituent distributions. A truly random variable has zero mean. Thus, in the case of a closed circuit containing a random noise source (e.g., a resistor) and a d-c source (a dry cell), the mean value of the current distribution function will be that due to the d-c source alone, whereas the mean squared variance about this mean will be that due to the noise source alone. If two or more independent random noise signals are added, the result will be a new random noise whose mean is the sum of the means of the original signals and whose mean squared value is the sum of the mean squared *values* of the original signals. This result follows from the additive nature of mean squared deviations of any superimposed probability functions, discussed above, since the mean squared value of a random noise (which has zero mean) is the same as its mean squared deviation.

This discussion of probability, probability distribution functions, and random noise, has been neither very thorough or very rigorous. It has not touched upon the large number of distribution functions, other than Gaussian, which are of interest in some cases. The reader who requires a practical, working knowledge of noise, probability, and information theory is referred to specialized texts on those subjects. For our purposes here, it is sufficient to recognize that many of the noise types encountered in electromagnetic information-transmission systems are, for practical purposes, random in nature. The amplitude-probability distribution of a random variable may be shown, by a statistical tool known as the *central limit theorem*, to be Gaussian. Hence, the Gaussian probability distribution is of great value in dealing with random noise arising from physical processes.

In addition to its amplitude-probability distribution, a noise signal is characterized by its spectral distribution, which is essentially the distribution of average noise power (mean squared amplitude) as a function of frequency. Noise whose frequency components are essentially constant ("flat") over a range of frequencies is said to be *white* over that band. It may be shown by means beyond the scope of this discussion that a signal that was truly random would have not only a true Gaussian amplitude probability distribution, but also a "white" spectral distribution from zero to infinite frequency. This condition follows intuitively from the definition of *random*, which requires that the amplitude at any instant be totally independent of that at any other instant. Thus the time between successive amplitude peaks or zero crossings of the random signal must be completely unbounded, which is equivalent to stating that all periods, and therefore all frequencies, between zero and infinity are equally probable. Since infinite band widths cannot exist in practical cases due to the limiting effects of parasitic reactances in physical devices, noise encountered in reality is pseudo-random at best. That is, it may approximate the mathematical characteristics of true randomness (Gaussian amplitude distribution, white spectral distribution) over a wide range, but ultimately it is restrained to finite peak amplitude and to finite band width. Noise arising from most physical processes tends to be white over a wide frequency range, so that the actual spectral noise distribution at the output of a filter network (receiver, etc.) is readily approximated from knowledge of the band-pass characteristics of the network. Examples of this approach will appear later in this chapter.

RECOVERY OF SIGNAL INFORMATION FROM A NOISE ENVIRONMENT

Assume that it is required that a pulse signal of the form shown in Fig. 2–4(a) be received and detected. The general shape and *RF* frequency of each pulse is known, but the periodic times of arrival of the pulses are known only within rough limits, and are spaced by several hundred times the pulse width.

Furthermore, a white Gaussian noise (represented by Fig. 3–4(B)) is superimposed on the desired signal at the receiver input. This would be the typical result of broad-band noise jamming. A conventional oscilloscope (presenting amplitude versus time) might show the composite signal somewhat as indicated in Fig. 3–4(C). It is apparent that, if the noise amplitude greatly exceeds the signal amplitude, it will be very difficult for an observer to detect and accurately locate the signal, i.e., the desired results of jamming. Detection and location are difficult, of course, because it is no longer a simple matter of establishing some threshold value and labeling any waveform that exceeds the threshold as a "signal." Because of the random nature of the noise (its essentially Gaussian amplitude probability distribution), it is known that there is some finite probability of its exceeding (at some instant) any practical threshold set. It is necessary to deal here with a phenomenon that may only be described in macroscopic terms; more specifically, the parameters of the probability distribution describing a given noise source as based on the average behavior of that source over a length of time that must be very long compared with any time intervals involved in the noise-generation process. It would seem intuitively reasonable that, if it is desired to arrive at some basis for distinguishing signal from noise, these descriptive parameters of the noise must be dealt with; consequently, it must be expected that the resolution process will require some finite amount of (averaging) time. One characteristic of probability distributions that is helpful was presented in the section on probability. It was stated, in essence, that the mean values of any two (or more) superimposed probability distributions are additive. Thus the mean value of the composite noise-plus-signal voltage envelope in the receiver should equal the mean value of the signal envelope alone, and should be zero during periods of no signal, since the mean value of a random noise signal is zero. To the extent that the averaging time is short compared with the periods of the lowest-frequency components present in the noise (which extend to zero frequency, at least theoretically, thus requiring infinite averaging time, i.e., true integration), the averaging process will fail to produce a

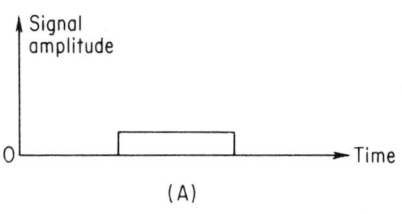
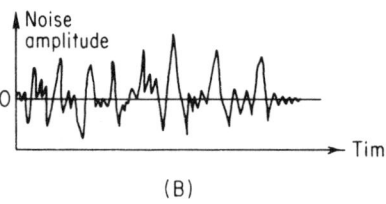
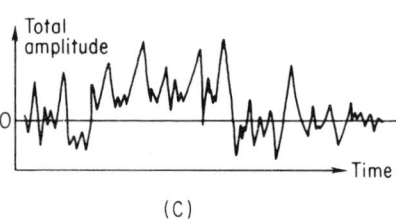

Figure 3-4. *Superposition of signal and noise.*

zero mean value for the noise. The degree to which the signal may be extracted from severe noise is, therefore, limited. A significant improvement in signal-to-noise ratio may be achieved in many practical cases, however, and the process is a very simple one.

As the signal pulse is made longer, the allowable averaging time is increased (provided pulse shape is not important), and the achievable improvement in signal-to-noise ratio (S/N) becomes greater. Theoretically, of course, an arbitrarily small, continuously running (CW) signal might be extracted from an arbitrarily large noise by averaging over an infinite period of time (true integration). Since the intelligence that may be conveyed by the knowledge that a signal exists is slight unless some parameter of the signal can be varied (modulation), it is apparent that in practice a compromise must be effected between information rate and signal-to-noise ratio.

The averaging process just discussed is in reality a simple low-pass filtering action; as the averaging time is increased, progressively lower-frequency components of the incident waveform are averaged (filtered) out, until the limiting case of infinite integration time is reached, corresponding to a low-pass filter of zero band width. Since the white noise to be contended with has a frequency spectrum that is essentially constant over all frequencies of interest, and since it of course possesses only a finite amount of total energy in practical cases (where its spectrum does not actually extend undiminished to infinite frequency), it is apparent that the noise energy in a given band of width Δf is directly proportional to the value of Δf. As Δf goes to zero, so does the noise power within that band. This is an intuitively obvious condition, since a zero band width signal corresponds to a pure sinusoid (of constant frequency) extending in time to positive and negative infinity, whereas a random noise, by definition, cannot possess any periodic component(s).

The desired signal, in practice, is never of zero band width. Information can only be conveyed via some sort of modulation of the carrier, resulting in a set of modulation products, or side bands, of non-zero width. Indeed, one of the most useful results from the field of information theory is the establishment of a direct relationship between system information-transmission capacity and band width. The usual variety of signal, however, has an overall band width considerably narrower than that of the noise of interest (exceptions may be found in some countermeasures situations). Nothing is added to signal intelligibility by utilizing a receiving band-pass wider than the signal band width; hence it is obvious that a band-pass just sufficient to enclose the entire signal will produce a greater signal-to-noise ratio than will a wider band-pass. It may even turn out in some cases that the band-pass that optimizes the S/N ratio is so narrow as to attenuate some of the signal side-bands, depending upon such variables as the shape of the passband, the nature of the received signal, and so on. (See Chapter 5.)

The receiving-filter "passband" has been discussed in rather vague terms, the implication being that the filter in question might possess a rectangular band-pass characteristic, i.e., that its response might be unity within the desired band and zero elsewhere. This condition, is, of course, a practical impossibility, although some existing types of filters do provide close approximations to this ideal. It should not be construed from these comments, however, that a rectangular passband is necessarily optimum. It will be seen in Chapter 5 that the choice of received signal-processing (filtering) techniques is a great deal more involved than the introductory discussion here would imply.

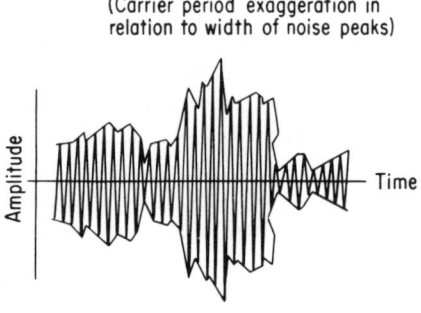

Figure 3-5. *Typical modulated-carrier noise-plus-signal wave before detection.*

In the discussion above concerning the averaging of a received waveform, it is the *envelope* of the signal-plus-noise composite (which carries the intelligence referred to) that must be averaged. The simple rectangular signal pulse, as well as the interfering noise, arrives at the receiving antenna in the form of a modulation envelope impressed upon an *RF* carrier. A great many cycles of the carrier frequency ordinarily correspond to the length of one signal pulse or one noise peak, as indicated in Fig. 3–5, hence the average value of the *RF* waveform is zero. The problem of *detection*, or extraction of the signal (video) envelope from the received wave, thus arises.

DETECTION AND CORRELATION

Probably the simplest method of detection consists of rectification of the *RF* or Intermediate Frequency (*IF*) signal, followed by averaging (low-pass filtering) over a sufficient time to effectively smooth out carrier-frequency ripple. This system displays two characteristics that make its use undesirable in some applications. It is shown in advanced texts on detection theory that such a system (which is called *incoherent* detection, for reasons that will become apparent in the next few paragraphs) has a deleterious effect on signal-to-noise ratio. The degradation is so small as to be insignificant in cases in which the detector input S/N is considerably larger than unity; it becomes increasingly severe for input S/N ratios approaching unity, as may happen in a severe noise-jamming environment. As the detector input S/N drops below unity, the incoherent detector further deteriorates the signal-to-noise ratio, so that a detector input S/N of -10 db yields an output S/N of -13 to -14 db.

The alternative to incoherent detection, as might be suspected, is known as *coherent* detection. In order to analyze the relative significances of, and

DETECTION AND CORRELATION   39

the differences between, coherent and incoherent detection, it is necessary to return to statistics and information theory. The term "coherent" as applied to signals in any discussion of communication theory may be defined as follows: if the descriptive parameters (i.e., frequency, phase) of two or more signals are functionally related in a specified manner, then those signals are said to be "coherent."

The existence or absence of such relationships between two functions of the same variable (in this case, time) may be investigated by resorting to the statistical concepts of *correlation*. Two time functions, $f_1(t)$ and $f_2(t)$, might be defined by the correlation factor ϕ_{12}, thus:

$$\phi_{12} = \lim_{T \to \infty} \frac{1}{2T} \int_{-T}^{T} f_1(t) f_2(t) \, dt. \tag{3-10}$$

Clearly, the correlation factor of two identical signals will be some positive number; that for two sine waves of the same frequency, but separated 90 degrees in phase, will be zero, since the product $f_1(t)f_2(t)$ will be positive as often as negative, averaging to zero over large T. Similarly, the average value of the product for any two functions arising independently, (e.g., a random noise and any other time function, or two sine waves of different frequencies) will be zero over a long time, T. If one of the two signals is delayed a time, τ, relative to the other, and the correlation factor is then determined as a function of the delay time, the yield is the cross-correlation function, $\phi_{12}(\tau)$, where:

$$\phi_{12}(\tau) = \lim_{T \to \infty} \frac{1}{2T} \int_{-T}^{T} f_1(t) f_2(t - \tau) \, dt. \tag{3-11}$$

Two unrelated signals will still yield $\phi_{12}(\tau) = 0$ for all τ, since τ merely introduces a varying "phase" relationship, and the concept of phase has no significance for independent signals. For the case of two signals with periodic components of the same frequency, however, $\phi_{12}(\tau)$ will be periodic in τ, yielding maxima for τ such that the signals are in a 0- or 180-degree phase relationship, and zero as τ passes through 90 degrees and 270 degrees.

The cross-correlation technique may be applied to the detection of a received signal in the presence of noise. Consider first the hypothetical arrangement of Fig. 3-6. An oscillator is (somehow) maintained at the exact frequency and phase of the expected received signal. When no signal is present, the oscillator signal is cross-correlated with received noise; the correlator output is zero because only signals arising from a common source can, in practice, be correlated over many cycles. When the received signal is present, the correlator output becomes a steady d-c voltage; thus there has been accomplished both signal detection and separation of signal from noise by a process of multiplication and averaging, using *a priori* knowledge of the received signal.

Full information regarding parameters of the received signal is seldom available. In many cases the local reference oscillator may be made to "track" the received signal (i.e., may be synchronized to it), using feedback circuitry to vary the oscillator's frequency and phase, (i.e., τ), so as to achieve maximum ϕ_{12} (really $\phi_{12}(f, \tau)$ in this case). Since no feedback loop is even theoretically capable of perfect tracking under all circumstances, the reference signal cannot be maintained in perfect correlation with the received signal. The multiplication-averaging process of coherent detection does not introduce the S/N ratio degradation inherent in incoherent detection (rectification-averaging), however, so that in many cases the type of detection system just described will provide a significant advantage in S/N over an incoherent

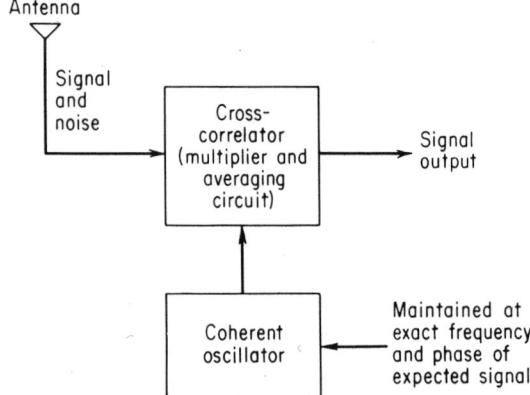

Figure 3-6. *Hypothetical (ideal) cross-correlation receiver.*

detector, despite degradation resulting from imperfect synchronization of the reference signal. In addition, the knowledge of received signal frequency and phase that is acquired in the process of deriving the coherent reference signal is useful in many applications.

Another statistical process closely akin to cross-correlation is autocorrelation. The autocorrelation function of $f_1(t)$ is $\phi_{11}(\tau)$, where

$$\phi_{11}(\tau) = \lim_{T \to \infty} \frac{1}{2T} \int_{-T}^{+T} f_1(t)f_1(t - \tau) \, dt. \tag{3-12}$$

Autocorrelation is thus seen to be equivalent to the cross-correlation of a function with itself (i.e., of two identical functions). By analogy with the latter case, it is evident that the autocorrelation function of any periodic component is periodic in τ; also $\phi_{11}(0)$ will be a positive number for any $f_1(t)$. (See Fig. 3-7(a).) A random noise is, by definition, independent in amplitude at any instant of its amplitude at any other instant; thus its $\phi_{11}(\tau)$ for any non-zero τ is zero (see Fig. 3-7(b)). Practical noise is never truly random, because it must pass through finite passbands in the antenna and receiver before reaching the detector, its high-frequency components are

DETECTION AND CORRELATION   41

attenuated, resulting in some degree of short-term autocorrelation and an autocorrelation function that decreases rapidly but not instantaneously with τ (see Fig. 3–7(c)). Because of this limitation, the autocorrelation detector is not capable in practice of extracting an infinitesimal signal from a large noise. Like the cross-correlator, it is capable of large improvements in S/N ratio when sufficient integrating (averaging) time is available. Unlike the cross-correlator, the autocorrelator requires no accurate *a priori* knowledge of the received signal. (See Fig. 3–7(d).)

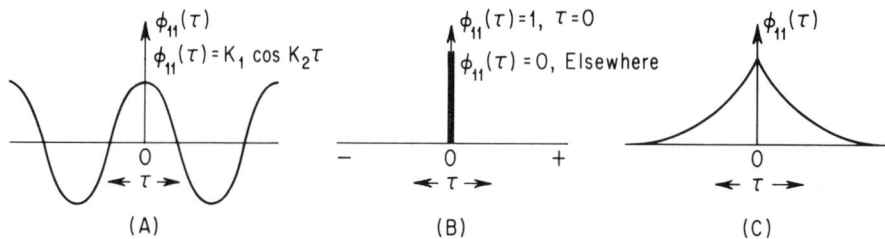

$\phi_{11}(\tau)$ *for sinusoidal signal* $\phi_{11}(\tau)$ *for random noise* $\phi_{11}(\tau)$ *for band-limited noise*

Autocorrelation functions, $\phi_{11}(\tau)$, for specific signals.

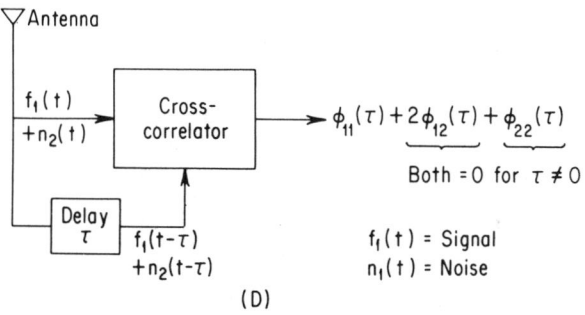

Simplified autocorrelation receiver.

Figure 3-7.

The correlation concept provides an extremely powerful and useful mathematical tool for dealing with many noise-and-signal problems. One useful result of the mathematical theory deserves further mention here because it provides rigorous justification of a statement that has heretofore been made on an essentially intuitive basis. It may be shown that the autocorrelation function of a given time function is the Fourier transform of that function's spectral distribution. Thus a truly "white" noise with spectral density constant over all frequencies has a "spike" autocorrelation function, possessing a finite value only for $\tau = 0$, and a zero value elsewhere (see Fig. 3–7(b)). Conversely, a pure sinusoidal signal extending to $\pm \infty$ in time, which has been noted previously has an autocorrelation function that is sinusoidal in τ, is proven to be a signal of zero band width (i.e., with a "spike" or "delta-function" type of spectral distribution) (see Fig. 3–7(a)). Similarly, it may readily be proven that an autocorrelator whose integration time approaches infinity is equivalent to a filter of band width approaching zero. It would appear, then, that the use of a very narrow filter in conjunction with an incoherent (rectifier) type of detector is no different from combining a long-integration time autocorrelator with a rectifier. The autocorrelator, for an appropriate value of τ, provides a detected output, however, and requires no additional rectification detection. Thus it is apparent that the S/N degradation inherent in incoherent detection must arise as a consequence of the non-linear* detection process (a rectifier being a highly non-linear element). It is common, therefore, to speak of linear detection and non-linear detection, the former referring to a process involving multiplication (so that the output is linearly related to the input), and the latter to one involving rectification (where the input-output relationship is decidedly non-linear). Although neither *linear* and *coherent* nor *non-linear* and *incoherent* are synonymous (they describe the detection process from different aspects), it is generally true in practice that a linear detector is also coherent, and a non-linear detector is not.

SIGNAL DETECTION IN THE PRESENCE OF ELECTRONIC COUNTER-MEASURES (ECM)

The theories thus far dealt with have been quite general, and no attempt has been made to relate them to the specific topic of electronic warfare or electronic countermeasures (ECM). In the application of electronic countermeasures, any of a large number of different waveforms may be injected into the enemy's information-carrying system for the purpose of reducing its ability to handle the desired information. All of these fall within the most general meaning of the term "noise" as used earlier in this chapter; it is convenient (and conventional) in order to avoid confusion, however, to be more specific in discussions of ECM, using the term "noise" to refer specifically to *random* noise or reasonable approximations thereto. Various forms of non-random (i.e., periodic) interference or jamming must be treated somewhat differently from random noise, both in mathematical analysis and in actual practice, and it is an aid to clarity to refer to these forms of ECM explicitly.

The problem of detecting a desired signal in an environment of random or quasi-random noise lends itself particularly well to mathematical treatment.

---

* Non-linear detection might alternately be referred to as *irreversible* detection, since the input signal could not be reconstructed from the output because of loss of information during detection. Linear detection, on the other hand, involves no loss of signal information and hence is essentially a reversible process.

SIGNAL DETECTION IN THE PRESENCE OF ECM

It will be instructive here to determine quantitatively the effect of noise on the probability of detection of the desired signal. In the simple case described earlier in this chapter, the decision as to the presence (or absence) of a signal at any instant is made by specifying a threshold, X_T; at any time when the received waveform exceeds X_T the decision is "signal present," when it is not, "no signal."

Considering only the instantaneous amplitude of the received waveform, (and not its polarity) the absolute amplitude probability distribution function of Gaussian noise is $p(|x|)$:

$$p(|x|) = \frac{\sqrt{2}}{\sqrt{\pi}\sigma} e^{-(|x|^2/2\sigma^2)}, \qquad 0 \leq x \leq +\infty. \qquad (3\text{--}13)$$

The coefficient is greater by a factor of 2 than that of the usual Gaussian normalized distribution because of the consideration of absolute values only; i.e., $p(|x_1|) = p(x_1) + p(-x_1) = 2p(x_1)$, since the distribution is symmetrical about $x = 0$ for random noise of zero mean.* If the signal, when present, possesses an amplitude X_S, then the probability that the signal-plus-noise waveform will exceed X_T when a signal is present, which is the probability of detection P_D, is equivalent by simple subtraction to the probability that the noise alone will exceed $(X_T - X_S)$. This equivalent is found by determining the area under the noise-amplitude probability distribution curve, between $x = (X_T - X_S)$ and ∞. Thus, the probability of detection of a steady signal of amplitude X_S in Gaussian noise as described above is

$$P_D = \int_{(X_T - X_S)}^{\infty} \sqrt{\frac{2}{\pi\sigma}} e^{-x^2/2\sigma^2} dx, \qquad X_T \geq X_S. \qquad (3\text{--}14)$$

If the threshold X_T is set lower than the signal amplitude X_S, the probability of detection is unity. P_D could not, of course, be improved beyond this point, which occurs for $X_T \leq X_S$, and it will be apparent from what follows that the probability of false alarm would be adversely affected by setting X_T smaller than X_S.

The probability of false alarm, P_F, is simply the probability that noise alone will exceed the threshold X_T, even when no signal is present. This probability is simply

$$P_F = \int_{|X_T|}^{\infty} \sqrt{\frac{2}{\pi\sigma}} e^{-x^2/2\sigma^2} dx. \qquad (3\text{--}15)$$

The Gaussian probability distribution function is readily integrated only by numerical means; however, solutions for specific ratios of the limits of

---

* Note that the factor of 2 might be omitted, consideration then being given to both + and − values of x, i.e., set "signal-decision" thresholds at $\pm X_T$. In the equations that follow, the integration limits would then reflect integration over symmetrical areas on both sides of the zero mean, with the same result arrived at here.

integration to σ are widely available in tabular form. Thus P_F may be determined for any specified values of X_T and σ. Further, specifying a value of X_S makes possible determination of P_D. The parameter X_T may be eliminated and families of curves of P_D versus X_S/σ, the signal-to-noise ratio may be plotted with P_F as the parameters. Fig. 3-8 is such a plot. S/N is the received signal-to-noise power ratio, equal to

$$\left(\frac{\frac{X_S}{\sqrt{2}}}{\sigma}\right)^2.$$

The factor of $\sqrt{2}$ is necessary since X_S as defined above is a peak amplitude (for a sinusoidal signal), whereas σ is the Root Mean Square (RMS) amplitude of the noise.

In order to illustrate the practical application of probability in a typical situation, consider a search radar that is required to detect a specific target at least 95% of the time, i.e., with $P_d = 0.95$. In addition, it is required that a false alarm be presented not more than once per 10 sec (comparison of results from one sweep to the next is used to prevent action from being taken against false targets). If the radar band width is 5.0 mcs, then approximately $2 \times 5.0 \times 10^6 = 10^7$ samples/sec are evaluated by the threshold device, according to elementary information theory. Thus one false alarm/10 sec corresponds to one false decision in 10^8, or a false-alarm probability, $P_F = 10^{-8}$. From Fig. 3-8, the required received signal-to-noise power ratio is $S/N = 15$ db approximately, or $31:1$. Thus the (peak) power of each received target-echo pulse must exceed the power level of the noise within the receiver band width by a factor of about $31:1$. It will be seen later how this technique may be extended to yield the characteristics required of a noise jammer in order to limit to any desired range the ability of the radar to detect the given target with a specified P_D and P_F.

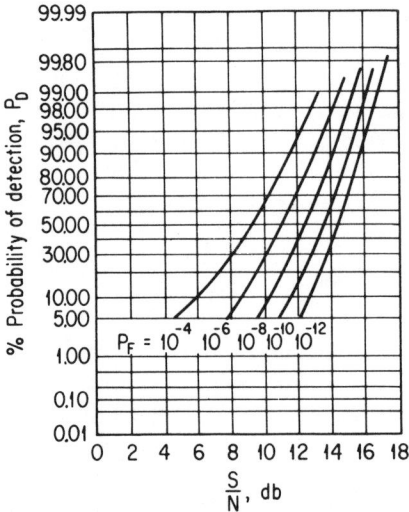

Figure 3-8. *Probability of detection of a signal for various false-alarm probabilities, P_F, vs. signal-noise ratio (Gaussian noise assumed).*

Thus the effect on a communications system of random noise, whatever its origin, may readily be determined. Jammers that transmit noise for the purpose of disrupting enemy systems are in wide use at present. In addition to this technique for introducing uncertainty into the signal-detection process, a second approach involves the injection into the enemy system of signals resembling (at least to a degree) the desired signal. The purpose of this approach is to cause confusion, saturation of signal-handling and data-processing equipment, or dilution of enemy defenses. In many cases the techniques required to minimize the effects of false signals on a system's information capacity are quite different from those employed in a random-noise environment. Similarly, the power, stability, and other requirements on the jammer are different for the two different types of ECM equipments.

The noise jammer is ordinarily required to provide a substantial advantage of noise power over signal power (i.e., a large Jam-to-Signal ratio, J/S) at the receiving antenna of an enemy system, if effective disruption of that system is to be accomplished. The use of sophisticated receiving techniques, including coherent detection, often results in S/N improvements of 10 to 15 db or more in the receiver, so that a considerable J/S is required at the antenna to reduce the ultimate receiver, S/N (or S/J, in this specific case) far enough to degrade P_D/P_F significantly. The power density at a receiving antenna due to jammer radiation is

$$W_J = \frac{P_J G_J}{4\pi R_{JR}^2}, \qquad (3\text{-}16)$$

where P_J is the jammer-transmitted power, G_J is the jammer antenna gain in the direction of the receiver, and R_{JR} is the range from jammer to receiving antenna. In many cases, it is desirable for the jammer to transmit a band of noise considerably wider than the band width of the victim system, in order to reduce the required frequency setting accuracy of the jammer or to allow jamming two or more systems on different frequencies simultaneously. If the victim receiver band width is B_R and if the total jammer power is spread uniformly over a band width of B_J, then the total jamming-power density within the frequency range that will enter the receiver passband and contribute to jamming effectiveness is

$$W_{JR} = \frac{P_J G_J B_R}{4\pi B_J R_{JR}^2}. \qquad (3\text{-}17)$$

Thus it is apparent that it is desirable to use as narrow a receiver passband, B_R, as possible without attenuating the desired signal; in addition, any technique incorporated in the communications system that will force the enemy to increase his jammer band width, B_J, will similarly aid in reducing W_{JR}. The power density of the desired signal corresponding to W_{JR} for the jammer is, for the case of a simple one-way transmission path,

$$W_{TR} = \frac{P_T G_T}{4\pi R_{TR}^2}, \qquad (3\text{-}18)$$

where P_T is the transmitted signal power (assumed to lie entirely within the receiver passband),

G_T is the transmitter antenna gain in the direction of the receiver,

and R_{TR} is the transmitter-to-receiver distance.

Now, the received J/S ratio, assuming that both signal and jamming arrive from the same direction and hence experience the same gain in the receiving antenna, is

$$\frac{J}{S} = \frac{W_{JR}}{W_{TR}} = \frac{P_J G_J B_R}{P_T G_T B_J} \frac{R_{TR}^2}{R_{JR}^2}. \qquad (3\text{–}19)$$

It will be instructive to consider an example illustrating how Eq. 3–19 may be combined with Fig. 3–8 (or Eqs. 3–14 and 3–15) to solve a practical problem in ECM. Suppose that a transmitter transmits a power $P_t = 100$ w, with an antenna gain $G_t = 20$ db $= 100$. The receiver band width is $B_R = 100$ cps. A noise jammer is located between the transmitter and receiver, at a distance $R_{JR} = 0.5\ R_{TR}$ from the receiver, where R_{TR} is the transmitter-to-receiver range. The jammer antenna gain toward the victim receiver is 3 db $= 2:1$, and jamming-power output is uniformly distributed over a band width $B_J = 100$ kc/sec. The transmitter is keyed by pulses of duration 0.1 sec, and it is desired to jam the receiver so that it is restricted to a probability of detection (of each pulse) of 10% if adjusted for one false alarm per minute. (B_R of 100 cycles/sec yields approximately 2×100 noise samples per second, so one false alarm per minute is one per $2 \times 100 \times 60 = 1/1.2 \times 10^4$, or $P_F = 8 \times 10^{-5}$.) From Fig. 3–8, this result requires $S/N \doteq 6$ db $= 4:1$. Since jamming is assumed to be the predominant source of receiver noise (in practice this assumption must be confirmed by calculation of S/N without jamming), this may be restated as $J/S = \frac{1}{4}$. From Eq. 3–19 then, the required jammer power is

$$P_J = \frac{J}{S} \times P_T \frac{G_T}{G_J} \frac{B_J}{B_R} \times \left(\frac{R_{JR}}{R_{TR}}\right)^2 = \frac{1}{4} \times 10^2 \times \frac{10^2}{2} \times \frac{10^5}{10^2} \times \left(\frac{1}{2}\right)^2$$

$$\doteq 3 \times 10^5\ \text{w}.$$

This is clearly an exorbitant jamming-power level, and the fault lies primarily in B_J and G_J. If a "spot" frequency jammer is employed so that B_J approaches B_R, then P_J is of the order of 300 w. A further reduction in P_J required would be allowed by raising G_J. For $G_J = G_T$, only 6 w suffices to reduce P_D to 10% for $P_F = 10^{-4}$, under the conditions otherwise specified. The above example does not consider further processing of the signal and noise to improve S/N. The 10 per sec signal pulse rate, for example, might in practice allow reduction of B_R to the order of 10 cps by a coherent detection scheme, in which case the required 6 db S/N at the receiver output would result from a S/N (or S/J) of -4 db ($J/S = 4$ db) in the pre-detection band width of 100 cps. If the jamming signal arrives from other than the direction of maximum receiving antenna gain, its effective power is correspondingly reduced. If the jammer attempts to jam more than one receiver simultaneously, the jamming antenna will be limited in possible gain or in the accuracy with which it may be "aimed" at any one target.

ECM systems that generate and radiate somewhat realistic (i.e., resembling the desired signal) confusion or saturation signals may not require the power advantage over the desired signal sometimes needed by noise jammers. In some cases it may be desirable to simulate the strength of the desired signal, as well as its modulation—or other characteristics. The reconnaissance requirements for successful operation of such systems are ordinarily more stringent than those for a noise jammer, frequently including data on such parameters of the signal as exact frequency, modulation or coding characteristics, amplitude, time of arrival (or phase), etc. The burden of complexity and specialization imposed on the confusion/deception jammer by the requirements discussed above is accompanied by cost and reliability problems that often tend to partially counteract the advantages (e.g., lower power requirements in some cases) of such systems. The general form of Eq. 3–19 is applicable to such systems, except that the confusion signal will ordinarily lie entirely within the receiver passband (since this type of jammer is not basically a broad-band device); hence the term (B_R/B_J) is omitted (i.e., is unity).

Regardless of the type of ECM radiation, certain general principles are immediately evident for the improvement of desired signal-detection probability. High transmitted power and antenna gain, a narrow system band width, and techniques forcing the enemy to broader jamming band widths are always desirable. In addition, the use of a sharp (narrow) antenna pattern on both transmitter (for high gain) and receiver (to reduce interference from jamming signals arriving from other than the direction of the transmitter) will always be helpful. Note that, when jamming does arrive from the direction of the desired signal, receiving antenna gain *per se* increases both jamming and signal equally, hence is of no advantage in an ECM environment (except for the directional effects mentioned above).

Countermeasures and counter-countermeasures developments are of course intimately interrelated, in that knowledge of ECM devices is extremely helpful, if not essential, to the development of effective ECCM techniques. Since intelligence information is necessarily incomplete at best, it is necessary to design information-transmission equipment and associated ECCM "fixes" on the basis of minimum susceptibility to the types of ECM most likely to be encountered. The incorporation of good general design features, intended to minimize system susceptibility to a variety of interference types, is generally preferable to the use of a large number of specialized circuits and techniques applicable to only one type of ECM interference reduction. Where certain advantages intrinsically accrue to the information-transmission system and to the ECM system, respectively, it is logical, and usually profitable, to capitalize on such inherent advantages and to minimize, in so far as possible, the advantage to the enemy ECM system.

An outstanding example of a natural advantage favoring an ECM system over the information-carrying system is found in jamming of radar equipment. The power density in the transmitted radar wave decreases as the square of the range from the radar. Upon reflection and re-radiation from a target, usually located near the ECM source, the radar signal's power density is again attenuated as the square of range, so that the received signal is proportional to the transmitted power divided by the fourth power of the radar-to-target range. The jammer output, on the other hand, travels only a one-way path and arrives at the radar with a power density proportional to the transmitted jamming power divided by only the square of the jammer-radar range (see Eq. 3–16). Thus, depending on the separation distances, the jammer may need to radiate only a small fraction of the radar's transmitted power in order for the jamming signal and the radar echo to arrive at the radar receiver with equal amplitudes. These relationships will be developed more fully, with analytical treatment, in Chapter 5. The jammer, however, is ordinarily denied complete advance information regarding the radar set to be jammed, such as, for example the exact design and parameters of its ECCM protective devices. Reconnaissance techniques tend to provide some information but, at present, exact and continuous knowledge of detailed parameters of the transmitted radar signal is available only at the radar; this fact provides the radar with an inherent advantage to somewhat offset the range advantage inherent to the ECM system.

In the design of a radar, communications system, or radio-control system (e.g., a missile-command guidance link), the choice of signal waveform to be transmitted exerts a significant influence over the degree to which the system may be made immune to enemy ECM activity. Most early radar systems, for example, were designed with little thought to confusion of enemy reconnaissance or protection against subsequent enemy jamming. Consequently, the radiated signals from these systems generally would be distinctive "signatures,"* clearly indicating the system mode(s) of operation, purpose, and to a certain extent the variety of countermeasure action likely to prove most profitable. The signals transmitted by early pulse radars might often be characterized by *RF* frequency, pulse-repetition rate, pulse length, and usually by some sort of periodic amplitude modulation on the signal. Each of these parameters yields fairly reliable information about the radar system originating the signal. Thus, it is apparent that the first logical step in extraction of signal information from interference originated by enemy ECM equipment is to disguise (code), in so far as practical, the nature of the information-carrying system. At the same time, it must be presumed that the enemy will always apply some sort of simple ECM (e.g., noise jamming), and may eventually obtain full information as to the design of any given system. It is

* A discussion of "signatures" is taken up in Chapter 4.

then imperative that the system be configured not only to deceive or confuse the enemy, but also to achieve optimum performance in spite of whatever ECM he may employ.

A number of techniques may be used in the selection of transmitted waveform to enhance the reliability of received information in a noise environment. From the earlier discussion of filtering and correlation detection, it is apparent that a long (in time duration) pulse will ordinarily be more readily detected in noise than will a short one. Similarly, redundance, or the transmission of a piece of information more than once, provides a higher probability of successful and accurate reception. For example, if P_D for a single pulse is P_{D_1}, then the probability of missing the first but detecting the second pulse is:

$$P_{D_2} = (1 - P_{D_1})P_{D_1}; \qquad (3\text{--}20)$$

the cumulative probability of detection for two pulses is:

$$P_{D_T} = P_{D_1} + (1 - P_{D_1})P_{D_1}, \qquad (3\text{--}21)$$

$$\therefore \ P_{D_T} = 2P_{D_1} - P_{D_1}^2 \qquad (3\text{--}22)$$

if $P_{D_1} = 0.8$, $P_{D_T} = 0.96$, etc. Both of these techniques are accompanied by reduction of the information-carrying capacity of the system because of the increase in time required for the transmission of each "message." The transmitted signal parameters, amplitude, frequency, and phase may be modulated, employing various coding schemes and corresponding decoding filters or networks at the receiver. And finally, though not perhaps best described as "selection of the transmitted waveform," increasing the transmitted power and antenna gain has already been mentioned as an obvious step toward enhancement of the received S/N ratio.

The choice of transmitted waveform, as discussed above, is ordinarily dictated, at least in part, by a desire to use at the receiver certain signal-processing techniques known to provide a relatively high degree of information recovery despite jamming. Conversely, the intended purpose (and allowable complexity, weight, and so on) of a system imposes restrictions on the transmitted waveform, so that the receiver must be designed to "make the most" of whatever waveform is selected as optimum. The effect of these constraints will be taken up in Chapter 7. It is difficult to specify signal-processing techniques that will be applicable to all systems. In general, however, the known parameters of the expected signal provide some basis for discriminating between signal and jamming noise or false signals. In some cases the received signal may be distinguished from false signals by the very fact that it is weaker, e.g., the ECM signal may be unreasonably strong compared with the expected legitimate signal. Thus the application of some sort of amplitude discrimination on the basis of minimum-threshold or peak-limiting may be in order. Known modulation (coding) characteristics of the expected signal often provide identification useful in signal extraction. The application of filtering or correlation techniques already discussed falls within this category, since the received signal is expected to exhibit a periodicity whose characteristics are at least approximately known.

A knowledge of the spectral or other characteristics of the interfering signals or noise may be combined with similar knowledge of the probable nature of the desired signal to provide a means for distinguishing them. An interesting example of this case approaches the situation mentioned at the beginning of this chapter, in which complete knowledge of the interfering wave is available, and therefore this wave may be subtracted from the composite required signal-plus-noise wave to yield signal only. "Static" due to discharges of atmospheric electricity, or lightning, is known to possess an *RF* spectrum extending over several decades; in addition, the "modulation envelope" resulting from the gross characteristics of individual lightning strokes is largely independent of the *RF* frequency. That is, a simple receiver with an incoherent detector may be tuned to a wide range of *RF* frequencies and receive essentially the same detected output due to atmospheric electrical discharges. If a desired signal is being interfered with by such atmospheric noise, this simple receiver may be set to a near-by frequency free of signals, and the noise envelope may be detected and subtracted from the signal-plus-noise output of the regular receiver. Such systems have been employed successfully in commercial practice, yielding a considerable improvement in S/N ratio of the presence of "static."

Finally, the direction of arrival of the desired signal often serves to distinguish it from undesired signals or noise. Thus the use of directional antennas, as previously discussed, automatically discriminates to some degree against interference arriving from other than the desired direction. This technique is of course common in radar, communications, and control-system applications, in which the antenna directivity and gain provide system enhancement even without the presence of intentional countermeasures.

Undoubtedly, a great number of other techniques might be used in the selection of a complete system configuration designed for maximum immunity to enemy electronic countermeasures. The underlying principles, however, remain the same:

1. Selection of a transmitted signal to give the enemy the least possible information from reconnaissance, compatible with the requirements of receiver-signal processing.

2. Selection of these processing techniques in the receiver to make the best use of identifying characteristics of the desired signal, while also making as much use as possible of the known characteristics of the interfering noise or signals.

3. In some cases the information received by two or more receivers, or derived by two or more complete systems, at different locations or utilizing different principles or parameters or operation, may be compared (or correlated) to provide useful discrimination between desired and undesired signals. This technique may perhaps be included under the heading of "redundancy."

4

ELECTRONIC RECONNAISSANCE

THE PURPOSE OF ELECTRONIC RECONNAISSANCE

Military reconnaissance is a function that is basic to warfare. Active warfare requires the intereaction between forces; those forces can therefore locate each other; or at least one can locate the other.* Similarly, electronic reconnaissance is also basic in the specific consideration of electronic warfare. In addition to enabling the location of the enemy prior to an actual engagement, it is a well-developed method of providing inputs for military intelligence. Warfare is generally carried on with incomplete information concerning the enemy's planning and operations. It is the function of reconnaissance to reduce the incompleteness of the information. The objective of reconnaissance is always to gain information concerning the enemy; it will be considered here in two broad classes. These two classes are defined roughly as a function of time between gathering and operational use. First, there is the reconnaissance required to locate the enemy so that he may be engaged in combat, i.e., determining the position of an enemy task force so that it may be attacked. This type might be called tactical reconnaissance, and is characterized by having a relatively short-range objective.

Second is the reconnaissance designed to observe the enemy forces

---

* This relationship is discussed analytically by H. Brackney in "Dynamics of Military Combat," *Operations Research*, Vol. 7, No. 1, January–February 1959. Mr. Brackney develops the relationship $\frac{dn}{dt} = -p_m r_m$; where $\frac{dn}{dt}$ is the loss of force n as a function of time when engaged by force m with probabilities of kill p_m and attacking rate r. Further, he shows that this attacking rate r is a function of the search or reconnaissance time, i.e., $r = 1/T_{sm} + T_{fm}$; where T_{sm} and T_{fm} are search and firing (salvo) times respectively of the m force.

(equipment as well as manpower) for strategic planning. This type is characterized by having a relatively long-range objective. It is this classification that is of major concern in a "cold war" and is therefore of particular interest.

Electronic reconnaissance acquires this desired information by monitoring the electromagnetic radiation from various types of enemy systems. In general, the intercepted signals are recorded and analyzed by some element of the military intelligence organization. It is this analysis that provides the long-range contribution of electronic reconnaissance. A thorough understanding of the relationships between the technical characteristics of intercepted signals; their location, movement, density, and activity can yield considerable insight into the enemy's general strategy and tactics. Electronic reconnaissance for this purpose provides a most interesting combination of military intelligence and communications engineering. As might be expected, the terminology from both fields appears in a discussion of the subject.

ELECTRONIC INTELLIGENCE

Before entering into a detailed discussion of some of the technical problems encountered in electronic reconnaissance it is necessary to understand in greater detail the kind of information that can be derived from this operation. Understanding the requirements and applications for the output information of a reconnaissance program will allow a greater appreciation of the systems problems discussed later.

Analysis of the data collected provides information that can be divided into two broad classes. These two classes, operational information and technical information, are discussed in detail below.

A. OPERATIONAL INFORMATION

1. Types of systems in use. It is desired here to establish what types of radars, communications systems, data links, fire-control systems, missile-guidance systems, navigation aids, early-warning systems, and so on, are operational in the field. To observe a specific system in use it is necessary to monitor its operational frequency with the proper type of receiving equipment at the proper time. The special receivers developed to provide this broad monitoring function are referred to as "ferret" receivers. The use for which a system is intended is somewhat defined by its radiation parameters, i.e., modulation, antenna characteristics, frequency, and power. These characteristics provide an electronic "signature" that helps to establish a connection between the radiated signal and possible equipment applications. These data, or at least part of them, are recorded on magnetic tape or film when an intercept is made by a ferret receiver. Generally, the location and time of the intercept are also recorded. When the data are analyzed and compared with known techniques, a reasonable indication of the intended use can be established.

2. Number of systems in use. Direct counts, as well as statistical sampling techniques, give an insight into the numbers of each type of equipment in use. The density of emitters in metropolitan areas, near airports, around military bases, and along the borders of a country will obviously be greater than in remote areas. Activity in remote locations generally implies a development operation of a higher security classification or the presence of an unknown base. As an example, knowledge of the number of a particular type of air-defense radars located around a major city (New York or Moscow) will give connecting information as to the number of missile batteries in the area.

The number of intercepts per unit time, $I(t)$, is proportional to the targets present, T_a, and the area searched, $A_s(t)$; $I(t) \propto A_s(t) \times T_a$. It is, of course, required that the target be operating during the period of "overflight" and that proper reconnaissance equipment be used to allow the intercept. Note that in the relation for intercepts per unit of search time the concept of area (or location) is introduced. Counting the number of active emitters is not sufficient. It is far more meaningful to state that there are three new long-range surveillance radars located around Stalingrad than to say there are three such radars located within Russia. Therefore the tabulating of emitters must be carried on with respect to a more precise geographical location. The accuracy with which the location can be specified will be referred to as geographical resolution. The resolution achievable affects the accuracy of the counting (two similar radars might be counted as one) and is a function of search rates, number and direction of reconnaissance flights, and the parameters of the surveillance equipment being used. This problem also has bearing on establishing enemy tactics and deployment, which is discussed next.

3. Tactics and deployment. Information about a country's present and planned strategy can be determined by observing the use made of its electronic weapon systems. In the intelligence field, much analysis is accomplished by a method referred to as "indicators." An activity indicator is, as the name implies, a change in the level of activity of some observed function. For example, shortly before an ICBM is launched at Cape Canaveral there is a considerable increase in the electromagnetic radiation in the area. Tracking and guidance radars must be checked, communications links tested, and telemetry circuits energized. An observer monitoring this wide radio spectrum from well outside the launch area can accurately predict the missile firing.

In a tactical situation, knowing where and when specific systems are being used can establish what is referred to as an Electronic Order of Battle (EOB). If this information is derived from the monitoring of radar systems only, it is sometimes referred to as a Radar Order of Battle (ROB). The EOB/ROB are constructed from the location and identification of the radiating elements used in support of major weapon systems. In addition, it is often possible to intercept messages handled by the communications systems of an enemy. A reconnaissance effort may be specifically designed for this function. Under battlefield conditions, when it is desirable to achieve fast response time with such a program, interpreters would be necessary to translate the intercept (generally stored on tape). This necessity points out a vital, and often overlooked, application of the linguist to an electronic reconnaissance program. The U.S. Air Force has an extensive program in the teaching of Russian to its airmen at Syracuse University. It can be assumed that some of these men would find application in the programs discussed above should the need ever arise.

This aid in determination of an enemy's strategy and tactics on either a long- or short-range basis is perhaps the most important application of electronic reconnaissance.

B. TECHNICAL DATA

1. Establishing the level of technical development. Enemy systems may, from time to time, employ advanced techniques or radically new modes of operation. Naturally, the inclusion of a new weapon system in a country's arsenal is not always announced to a potential opponent. These new systems may include an extension of the communications frequencies, a new type of modulation, or a new mode of antenna scanning and pattern generation. These indicators and many others may represent a new weapon or a new application of an old weapon in the field.

It is important that the reconnaissance program be designed in a sufficiently general way that it will accept the unexpected information connected with these new signals. This requirement, of course, places a serious design complication on a ferret receiver in that it must be able to accept many different types of signals. In practice it is not possible to design one receiver to accomplish the entire task and therefore there are many types of equipment in use. Some ferret systems are designed to receive specific signals with great detail, i.e., a high degree of frequency resolution and accurate counting of the *PRF*. Other ferret systems are intended to explore, in a gross way, new sections of the frequency spectrum heretofore unused for military applications.

It is far more desirable to identify the presence of a new technique as it is being developed in the laboratory and testing stages than to do so when it becomes operational in the field. Detecting the new development in its early stages allows the necessary time for proper analysis and the taking of steps to counter its expected effect as it becomes operational. Naturally, since the emissions of this new signal are rather infrequent during its development and may be deep within a country's interior (and therefore not too accessible), the reconnaissance program is presented with additional problems. One possible solution might be to use a satellite as the reconnaissance vehicle. It would provide frequent passes over the generally inaccessible territory in question. However, this method has the disadvantage of making it difficult to modify the "airborne" equipment at will to accommodate the changing electronic environment.

Regardless of the method of acquisition, the important point is that a reconnaissance program must be able to recognize new and unidentified technical parameters as early as possible in their development since they are positive indicators of new weapon systems.

2. Determination of ECCM techniques. One of the important purposes of establishing the existence of new electronic techniques is to enable ECCM methods to be developed where necessary. For example, penetrating bombers often use mapping radar as a navigation aid. Should the enemy develop a missile designed to home on this type of mapping radar signal, the results could be fatal to naïve bombers using such a system. Modification of the signal characteristics might provide an effective form of ECCM.

Another procedure for gaining information on which to base ECCM development would be to activate elements of a nation's defense intentionally and carefully observe the ensuing activity. This stratagem could be accomplished by directing an aircraft, or flight of aircraft, to approach the early-warning radars of the enemy's defense system. How deep the penetration would be allowed to progress is a matter of conjecture and policy. Imagine the activity that would be created when, and if, a flight of unidentified aircraft should start approaching our DEW line radars. If these aircraft were fast reconnaissance units, carrying not bombs but thousands of pounds of reconnaissance and surveillance equipment, a large amount of information could be accumulated as a defense-activation procedure was followed out. Of course this dangerous game can be played by both sides. It can logically be carried one step further in that some of the equipment carried by the aircraft can include a complement of jamming transmitters. In this case the radars are intentionally jammed during the approach and the ensuing countermeasures (frequency shift, increased power, two radars working together to provide triangulation, and so on) are carefully recorded.

These procedures are both militarily and politically dangerous and are not generally subscribed to by major powers in time of peace. However, in a discussion of the problems of acquiring information from behind an "electronic curtain" it is appropriate to mention these methods.

The problem of ECCM, and ECCCM, and finally EC^nM can easily be carried to a ridiculous extreme. In the face of a jamming signal the ECCM used to restore normal operation is referred to as a "fix." Since new ECM techniques are continually being developed to jam existing electronic weapon systems, ECCM fixes are often "add-on" devices. ECCM fixes may range from a new side-lobe suppressed antenna for a radar to a simple R-C time constant added to the *IF* of a communications receiver. In either case, the important requirement is a careful evaluation of the technical characteristic of the hostile environment so that effective ECCM fixes can be developed.

The foregoing material has covered many of the requirements placed on an electronic reconnaissance program. It should be understood that outputs from this effort do not always have to stand alone and can receive supporting information from other sources such as photographic reconnaissance. However, much of the desired information, and particularly technical parameters of the signals, can best be acquired by the methods discussed here (with due allowance made for covert agents).

The remaining sections of this chapter will be devoted to a discussion of some of the more technical problems inherent in meeting the requirements for information considered above.

PROBABILITY OF DETECTING, PROBABILITY OF INTERCEPTING

In all the foregoing discussion the acquiring of information depended on the interception of the desired signal. The probability of intercepting this signal on a given mission is of course not a certainty (probability = 1) and is in fact a function of many time-varying parameters. For the sake of discussion, some of these parameters will be greatly simplified or ignored. Even then, it is necessary that the reconnaissance aircraft (or satellite) be in range during a period when the desired emitter is radiating, that the correct frequency and demodulation system be on hand in the vehicle, and that while it is within range the reconnaissance antennas "look" in the correct direction.

One of the first characteristics to consider is the probability that a desired emitter is operating during the period when observation is possible. In this discussion it will be assumed that the emitter is using an omni-directional antenna* and can be treated as a point source. It will further be assumed that the operation of the equipment is rather periodic in nature, having a regularly established operating time, t_1, and off time t_2. A reconnaissance vehicle can observe a point along its flight path for a time t. This observational time t is controlled by factors to be discussed shortly. A probability of the desired signal being on at any given time T can now be given as:

$$P_o = \frac{t_1}{t_1 + t_2}. \qquad (4\text{--}1)$$

Also, the probability of observing the signal by continuous looking during the time t is a function of

$$\int_o^t \gamma \, dt. \qquad (4\text{--}2)$$

Here, γ is a function of such parameters as a blip-scan ratio, signal-to-noise ratio, and so on.

Equation 4–2 has application indicated by the following derivation. For there to be no observation during the interval $t + dt$, observation must fail

* In Chapter 6 we shall treat this problem again, allowing the antenna to be directional and scanning.

during both of the intervals, t and dt. Let the probability of not observing during $t + dt$ be $\lambda(t + dt)$; the probability of not observing during the period t is $\lambda(t)$; and for the period dt is $1 - \gamma\, dt$.

These relationships, assuming independence, can be expressed as

$$\lambda(t + dt) = \lambda(t)(1 - \gamma\, dt)$$

which yields the differential equation

$$\frac{d\lambda(t)}{dt} = -\gamma\lambda(t). \tag{4-3}$$

Rearranging and integrating,

$$\int \frac{d\lambda(t)}{\lambda(t)} = -\int \gamma\, dt. \tag{4-4}$$

gives

$$\ln \lambda(t) = -\gamma t; \quad \text{or} \quad \lambda(t) = e^{-\gamma t}. \tag{4-5}$$

Therefore, the probability of observing during the time period t can be written as

$$P_{ob} = 1 - e^{-\gamma t}. \tag{4-6}$$

Although a time t is devoted to "looking," the probability of the signal being present (on or not on) must also be included.

Considering the results of Eq. 4–1, the probability of not detecting, P'_d becomes $1 - P_o P_{ob}$, or in terms of the definitions:

$$P'_d = 1 - \left[\frac{t_1}{t_1 + t_2}(1 - e^{-\gamma t})\right]. \tag{4-7}$$

Equation 4–7 gives the probability of not detecting during a looking time t. For $t = 0$, $P'_d = 1$, which indicates that if no time is spent looking (while within range), not detecting is a certainty. Also, if the on time $t_1 = 0$, $P'_d = 1$, which indicates that if the emitter is off, it obviously cannot be detected.

The probability of detecting, P_d, under these conditions is:

$$P_d = \left[\frac{t_1}{t_1 + t_2}(1 - e^{-\gamma t})\right]. \tag{4-8}$$

Equation 4–8 gives the probability of detecting an emitter, operating as defined above, and assuming the reconnaissance vehicle will be within range during the time t.

SEARCHING MODES

The area exposed to search in a given interval of time is related to the velocity of the aircraft, its altitude, and the type of antenna system being used for the search. A combination of these factors in two types of searching efforts will be considered. The first class might be called a downward-searching mode, or over-flight. In this case the vehicle, either a satellite or a very high-flying aircraft scans a swath of earth with a downward-looking antenna pattern.

Two examples are shown in Fig. 4–1. The second class might be called a horizontal mode and is shown in Fig. 4–2. The method shown in Fig. 4–1(B) is primarily restricted to wartime use since it requires flying over the points

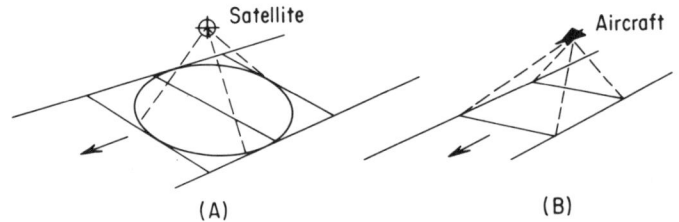

Figure 4-1. *Typical overflight antenna scanning patterns.*

of interest. The airspace required for the reconnaissance shown in Mode (A) of Fig. 4–1 and in Fig. 4–2 is unrestricted and open to use as desired. As might be expected, the method of Fig. 4–1(B) has certain advantages over the other two and will be discussed first.

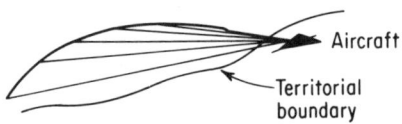

Figure 4-2. *Horizontal scanning pattern from within friendly territory.*

Consider now the type of searching shown in Fig. 4–1. The search is being carried on from a height h, and the vehicle is traveling with a velocity v. The downward-searching antenna has a look-angle θ, as measured at the half-power points. These conditions are shown in Fig. 4–3.

The rate of area search is:

$$\frac{dA_s}{dt} = 2vh \tan \frac{\theta}{2}. \quad (4\text{–}9)$$

For a given velocity v, and altitude h, of flight, the search rate is directly related to θ. The geographical resolution, however, decreases as θ increases. When an intercept is made its position is recorded in relation to the time of flight and other basic navigational data.

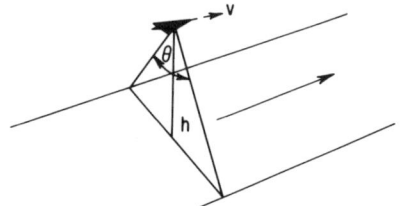

Figure 4-3. *Rate of area scan as a function of h, θ, and v.*

The area of uncertainty in geographical determination is equal to the area illuminated at the time of intercept. For the two conditions shown in Fig. 4–1 it is:

$$\text{for a circular antenna pattern} \quad A = \pi \left(h \tan \frac{\theta}{2} \right)^2 \quad (4\text{–}10)$$

$$\text{for a square antenna pattern} \quad A = \left(2h \tan \frac{\theta}{2} \right)^2. \quad (4\text{–}11)$$

For a single intercept the resolution can only be defined as lying within the illuminated area as given by Eqs. 4–10 or 4–11 for the two cases shown here. If additional intercepts of the same signal are made on different flight paths (or at a later time) a finer geographical resolution can be obtained. An example of this condition is shown in Fig. 4–4. In (A), the intercept taken at I_1 indicates an emitter within the shaded area. In (B), two intercepts taken of the same emitter at intercept points I_1 and I_2 limit the possible location to the much smaller shaded area shown.

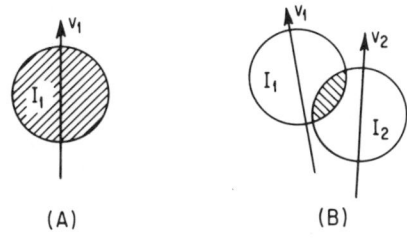

Figure 4-4. (A) *Illuminated area after a single intercept.* (B) *Improvement in emitter location due to two intercepts.*

When the intercepts are recorded in X and Y coordinates (or any convenient navigational system), the enhancement in resolution provided by the second intercept can easily be determined.

If

$$I_1 = (X_1, Y_1)$$
$$I_2 = (X_2, Y_2)$$

an overlap (shaded area) exists when

$$\overline{I_1 I_2} = \sqrt{(X_1 - X_2)^2 + (Y_1 - Y_2)^2} < 2R \qquad (4\text{--}12)$$

where $R = h \tan \theta/2$ from Eq. 4–10 and $\overline{I_1 I_2}$ is the linear distance between the points of the I_1 and I_2 intercepts.

The magnitude of the area included in the overlapping section is:

$$A_{OL} = \left[\frac{\pi R^2 \cos^{-1}(\overline{I_1 I_2}/2R)}{90} - \overline{I_1 I_2} R \sqrt{1 - \left(\frac{\overline{I_1 I_2}}{2R}\right)^2} \right] \qquad (4\text{--}13)$$

and the resolution improvement factor, K, can be expressed as Eq. 4–10 divided by Eq. 4–13.

Therefore
$$K = \frac{\pi (h \tan \theta/2)^2}{A_{OL}}. \qquad (4\text{--}14)$$

These considerations have particular interest in reconnaissance systems in which the X and Y coordinates can be stored in a computer and compared with a new set of intercept data at a later time. A calculation can then be made to establish the location improvement that has been accomplished by the second set of intercept data.

PROBABILITY OF OVERFLIGHT

Antennas used as shown in Fig. 4–1 can obviously search only a limited area in a given time. This rate of search was given in Eq. 4–9. Consider now that some unit of time t is equivalent to a single reconnaissance flight n. For example, if an aircraft has three hours of flying time available before it must land for logistic reasons (fuel, and so on) that is a single flight, or $n = 1$. If this same aircraft is used for 12 hours of reconnaissance flying, n must equal 4. In the case of a satellite traveling at an altitude that allows, say, 20 minutes of useful reconnaissance time out of each complete orbit, then n will be defined as equal to 1 for each such 20-minute pass. For 10 complete orbits, n equals 10, and the total reconnaissance time is therefore 200 minutes.

With this limited rate of search there exists a probability, or condition, whether or not on a given flight, the target (emitter) will fall within the search pattern. This problem can be defined by stating that it is required that some large total area A be searched, in order to locate a specific emitter. This emitter has equal probability of being anywhere within area A. An area A_s is searched on each of n flights. It is assumed that the searching will be done by linear scanning without overlap (as a farmer plows a field), and further that the emitter does not move during n successive flights.

Since on each flight (A_s/A) of the total area is searched and n flights are made, the probability of flying over a target, i.e., overflight, can be written

$$P_{ov} = \min\left[n\left(\frac{A_s}{A}\right); 1\right] \tag{4-15}$$

where A_s is given by the integration of Eq. 4–9

$$A_s = \int_0^{t_f} 2vh \tan\frac{\theta}{2} \, dt_f. \tag{4-16}$$

The limit t_f represents the flight time involved and $n = f(t_f)$ as indicated above.

The probability of overflight (Eq. 4–15) combined with the probability of detection (Eq. 4–8) provides a relationship for the total probability of intercept, P_I, with consideration of the parameters as discussed here:

$$\therefore P_I = P_d P_{ov} \tag{4-17}$$

$$P_I = \frac{t_1}{t_1 + t_2}(1 - e^{-t\gamma})\left(\frac{n}{A}\int_0^{t_f} 2vh \tan\frac{\theta}{2} \, dt_f\right). \tag{4-18}$$

Not all targets are fixed in position during a search period. Some emitters, such as a radar van, have mobility, or rate of travel, that is slow relative to

PROBABILITY OF OVERFLIGHT   61

any flight n but that may be fast (cover considerable distance) compared with the time between successive flights.

This relationship can be illustrated by letting area A be scanned in four flights; n_1, n_2, n_3, and n_4, as shown in Fig. 4–5. However, in the interval between flights n_1 and n_2 let the target T move, as shown, into the previously scanned area A_{s_1}. In this case the entire area can be completely searched without accomplishing overflight of the target. With the target now allowed to move in an unknown fashion with respect to the search pattern the two become unrelated events. When an area A_s is searched on a single flight the probability of no overflight is

$$P'_{OV} = 1 - \frac{A_s}{A}. \qquad (4\text{–}19)$$

Also to be considered is the condition that, should overflight be accomplished, the emitter may not be on (or may not be detected) during the overflight period. From Eq. 4–8 this condition gives:

$$P'_d = 1 - \left[\frac{t_1}{t_1 + t_2}(1 - e^{-t\gamma})\right], \qquad (4\text{–}20)$$

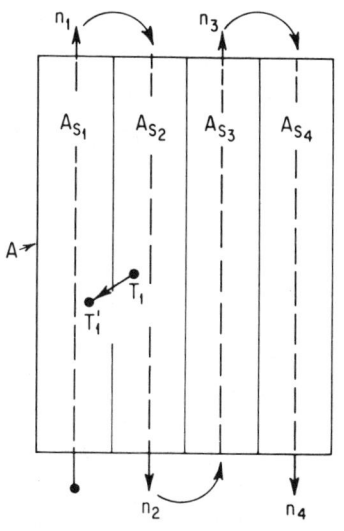

Figure 4-5. Linear scanning of area, A, in four consecutive flights, n_1, n_2, n_3, n_4. It is allowable for the target to move from T_1 to T_1', as shown, during the searching process.

which, combined with Eq. 4–15 for a single flight, yields

$$\frac{A_s}{A}\left[1 - \frac{t_1}{t_1 + t_2}(1 - e^{-t\gamma})\right]. \qquad (4\text{–}21)$$

Adding Eq. 4–19 to Eq. 4–21 and considering for n flights gives the probability of not intercepting the target as follows:

$$1 - P_I = \left[\underbrace{\frac{A_s}{A}\left(1 - \frac{t_1}{t_1 + t_2}(1 - e^{-t\gamma})\right)}_{\text{Prob. of flying over but not detecting}} + \underbrace{\left(1 - \frac{A_s}{A}\right)}_{\substack{\text{Prob. of not flying} \\ \text{over on a single flight}}}\right]^{n \text{ (flights)}} \qquad (4\text{–}22)$$

Simplifying and solving for the probability of interception P_I, for the moving target condition yields:

$$P_I = 1 - \left[1 - \frac{t_1}{t_1 + t_2}(1 - e^{-t\gamma})\frac{A_s}{A}\right]^n. \qquad (4\text{–}23)$$

This discussion has dealt with some of the requirements for establishing an initial intercept. The fact that the intercept has been accomplished indicates that the signal has been received and recorded. This result in itself provides considerable information relating to the technical parameters of the signal (frequency, *PRF*, *PW*, and so on). It is also obvious that some gross definition of the emitter's location has been established by virtue of the initial intercept. In some circumstances, as discussed under the topic of intelligence requirements, establishing the existence of the emitter and a gross geographical location is quite adequate. However, if the objective is to reconstruct carefully the deployment of enemy forces and supply detailed information for an *EOB*, a finer geographical resolution may be desirable. Improvement of geographical resolution was introduced briefly in connection with Fig. 4–4. There are many patterns and scanning modes that might be used to enhance the emitter resolution. It is now appropriate to consider some of the relationships that are encountered in various approaches to the problem.

OVERFLIGHT RESOLUTION

First, the overflight scanning pattern shown in Fig. 4–1(B) will be discussed. During the time of search it will be assumed that the emitter is continually radiating, and can be detected, i.e., $P_I = 1$. The object here is to locate given emitters with some specified accuracy. It is now desirable to define this problem so that relationships may be derived that allow comparison between such parameters as flight time required, resolution obtainable, and antenna-aperture effects.

The total area to be scanned is an area A, which is L miles long and W miles wide. The antenna used in this search will have an effective pattern S miles on a side (effective area therefore S^2 miles). The area will be scanned in strips, as shown in Fig. 4–5, until the desired intercept to be resolved is made. When an initial intercept is made the area of resolution is S^2 (see Eq. 4–11). This now becomes the area in which additional scanning is required if the resolution is to be improved. The area will be scanned in halved strips as shown in Fig. 4–6(B). The area of resolution, a, after n flights across the area S^2 is

$$a = \frac{S^2}{2^{n-1}}, \quad (4\text{--}24)$$

or

$$n - 1 = \frac{\ln(S^2/a)}{\ln 2}, \quad (4\text{--}25)$$

which gives the number of scans devoted to reducing the initial intercepted area S^2.

Since each of these passes need only be S miles long, this flight track length T' is

$$T' = S(n - 1), \quad (4\text{--}26)$$

and the flight-track length to scan the entire area A to produce initial intercept is:

$$T = \frac{WL}{S}. \tag{4-27}$$

∴ the total track length becomes:

$$T_T = T + T' \tag{4-28}$$

substituting
$$T_T = \frac{WL}{S} + S\left(\frac{\ln(S^2/a)}{\ln 2}\right). \tag{4-29}$$

The total track length required to produce a desired resolution is dependent on the antenna pattern used. The larger the area enclosed in the pattern

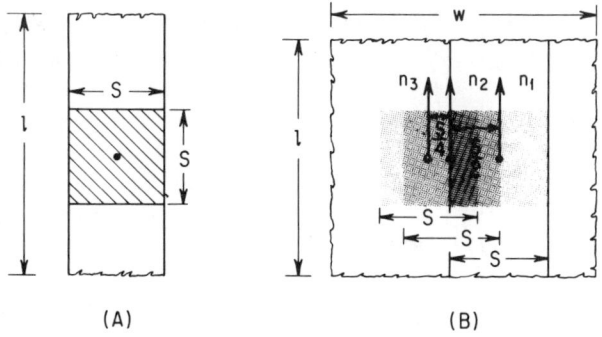

Figure 4-6. (A) *Area of a scanned strip illuminated by an antenna pattern of S units on a side.* (B) *An initial area of S^2 units being rescanned by flights n_2 and n_3. Each consecutive flight overlaps half of the remaining area defined by the initial intercept n_1.*

(S^2), the smaller the $T = WL/S$ section of the flight, but the more passes required to reduce the final resolution to the desired value.

This value in total track (T_T) length as a function of antenna-aperture opening S, can be minimized by taking the derivative of Eq. 4-29.

$$\therefore \frac{dT_T}{dS} = -\frac{WL}{S^2} + \frac{1}{\ln 2}\left[2 + \ln \frac{S^2}{a}\right]. \tag{4-30}$$

Setting equal to zero and solving yields:

$$WL \ln 2 = S^2\left[\ln \frac{S^2}{a} + 2\right] \tag{4-31}$$

rearranging
$$WL = 1.45S^2\left[\ln \frac{S^2}{a} + 2\right]. \tag{4-32}$$

Realizing that WL equals the area A, the minimum track T_T results for a desired area of resolution, a, when

$$A = 1.45S^2\left[\ln\frac{S^2}{a} + 2\right]. \tag{4-33}$$

Shown in Fig. 4-7 is a plot of S as a function of A for two levels of resolution a.

Figure 4-7 shows that a minimum T_T for the desired area of resolution can be obtained with an effective ground-scanning coverage of $S = 3.6$ miles on a side in this example. Equation 4-11 relates this scanned area (S^2) to the altitude of flight and the half-power "look" angle of the antenna used in search. Through these relationships, the height of search (h) can be controlled to produce the results indicated by Fig. 4-7 for the desired resolution and a given antenna. With h specified, and v set by the aircraft used, the total search time can be determined by Eq. 4-9.

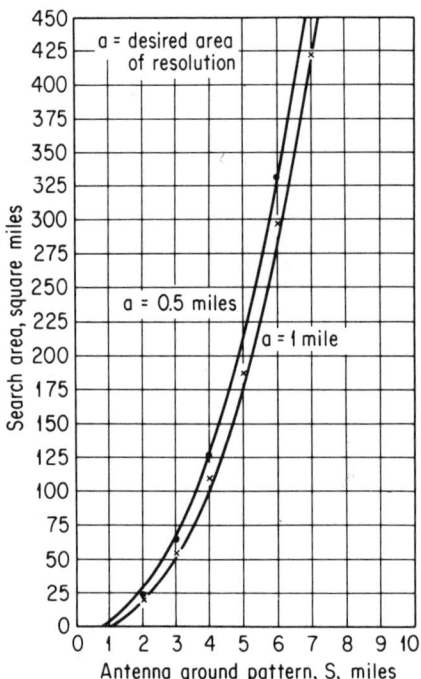

Figure 4-7. *Antenna ground pattern, S, as a function of total search area, A, as related by Eq. 4-33.*

To clarify this case a typical problem will be discussed. Consider an Army reconnaissance drone having a top speed of 250 mph and a maximum cruising time of 30 minutes. This drone is equipped with a downward-looking antenna having a half-power beam width of 120 degrees. It is desired to provide reconnaissance over 300 sq miles and to locate an enemy transmitter known to be within this area to an accuracy of 1 mile. This objective should be accomplished by selecting a flight plan that requires a minimum total flight time (minimum T_T). It is obvious that the maximum drone capability is a track length of 250 mph × 0.5 hours = 125 miles, and that the flight plan selected should not exceed this limit if the mission is to be accomplished with a single sortie. The area scanned is a function of the drone's altitude. To establish the optimum altitude Eq. 4-32 must be solved for the resolution of $a = 1$ sq mile. From the plot shown in Fig. 4-7 it can be seen that for a total area of 300 sq miles a pattern approximately 6 miles wide should be used.

Inserting $S = 6$ miles, $WL = 300$ mile^2 and $a = 1$ mile^2 into Eq. 4–29 gives:

$$T_T = \frac{300}{6} + 6\left[\frac{\ln 36}{\ln 2}\right] = 50 + 31 = 81 \text{ miles.}$$

The first factor (50 miles) is the track length to search the entire area and the second factor (31 miles) is the additional track length used to reduce the intercept to the desired resolution. The track length of 81 miles is within the capability of a single drone; hence the maximum time necessary to locate the target transmitter would be

$$\tfrac{81}{125} \times 0.5 = 19.4 \text{ minutes of flying time.}$$

The altitude of flight (see Fig. 4–3) necessary to generate $S = 6$ miles for the antenna characteristics given is

$$h = \frac{S}{2 \tan \frac{120}{2}} = 9100 \text{ ft.}$$

This formula defines the necessary parameters for the drone mission. As a comparison, consider a flight profile that produces a resolution of 1 sq mile directly. This profile can be accomplished by flying at an altitude of 1530 ft. The ground "swath" with the drone antenna given will be 1 mile, and the total area capable of being searched in a single sortie is 125 sq miles. Under these conditions it will require 2.4 sorties ($T_T = 300$ miles) or 72 minutes, to cover the entire 300 sq miles. The track length, and hence the search time, would also be greater if an altitude greater than 9100 ft were chosen. For example, an altitude of 22,800 ft produces a ground pattern of $S = 10$ miles. For this condition $T_T = 97.5$ miles and would require 23.5 minutes to produce the desired intercept resolution of 1 mile. Hence, this approach provides an optimum tactic when it is desired to locate a known target with a defined resolution within a given area.

"HORIZONTAL" RESOLUTION

Overflight as shown in Fig. 4–1 and discussed above is not always possible. During a time of peace, political constraints will not openly permit the violation of a country's "airspace" rights. Basically it is this overflight function that has been rejected by the non-acceptance of open-skies policies.

A mode of reconnaissance that does not violate any of the above-mentioned constraints is shown in Fig. 4–2. In this mode of operation the flight path is restricted so that the target always lies to one side of the aircraft. This attitude requires that the antenna "look" horizontally (rather than vertically as before) out of the side of the aircraft to effect an intercept.

Consider now the problem of defining geographical resolution for an intercepted signal in a horizontal search mode similar to that shown in Fig. 4–2. Taking the emitter to be a point source, it is desired to determine its location as closely as possible. One typical method of establishing its location is to take intercepts from two different points along a known baseline and calculate its position by triangulation. This approach introduces an area of uncertainty because of the finite beam width of antennas as measured at the half-power points. This area of uncertainty theoretically varies from infinity to some minimum as a function of the intercept angles. It is desired to see how this area varies as a function of these angles and to determine the relationship of these angles to the flight path that provides the minimum area of uncertainty, i.e., the highest geographical resolution.

A minimum of two intercepts is required to define the target location in this mode of operation. Because of the acceptance of this method of reconnaissance, the relationship between the positions of these two intercepts in

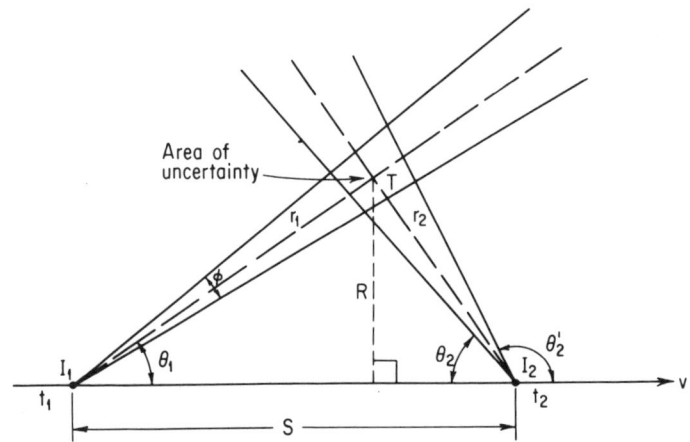

Figure 4-8. *Representation of two intercepts taken of emitter, T, from points I_1 and I_2 along a common flight path, S.*

defining target location (reducing the area of uncertainty) will be discussed in some detail.

To define the problem, assume that the reconnaissance aircraft is flying along a flight path S, a horizontal distance r, from the true target position T. The minimum distance between the target and the flight path is R. The aircraft is equipped with either one scanning horizontal searching antenna or two fixed ones. The question can now be asked, "If the first intercept, I_1, at time t_1, is made at a bearing θ_1, relative to the flight path, what is the optimum bearing θ_2, to establish the second intercept I_2 so as to reduce to a minimum the intercepted area of uncertainty?" The geometry of the problem is shown in Fig. 4-8. The area of uncertainty can now be visualized at the two extremes. First, assume that the intercept I_2 was taken at a time $t_2 = t_1 + dt$ as dt approaches zero. The target position can then be defined only at a nominal angle θ_1 (relative to the aircraft), and a range r_1; where $0 < r_1 < \infty$.

"HORIZONTAL" RESOLUTION

Allowing I_2 to be taken at $t_2 = \infty$ provides no additional information since the beam width is also infinity in the target area under this condition. Thus, the position of the target cannot be defined any better by two intercepts than by one if the second is taken from extreme points. It is desired to determine if an optimum point exists between these limits, and if so, what the criterion for establishing this point is.

In Fig. 4–8 the following elements are defined for reference:

- T the true target position,
- S path of flight; assumed straight during period of interest,
- I_1 position of first intercept,
- t_1 time of first intercept,
- θ_1 bearing to target at first intercept,
- r_1 range from I_1 to T,
- R minimum range from flight path (S) to target,
- v velocity of aircraft,
- ϕ effective antenna beam width at the half-power points.

In like manner the subscript 2 is used to denote functions of the second intercept (I_2).

The airborne reconnaissance antenna has an effective beam width angle at the half-power points of ϕ. Also, the same antenna, or a similar one, will be used for both intercepts. The width of the beam at any distance r is therefore:

$$\rho = 2r \tan\left(\frac{\phi}{2}\right). \quad (4\text{--}34)$$

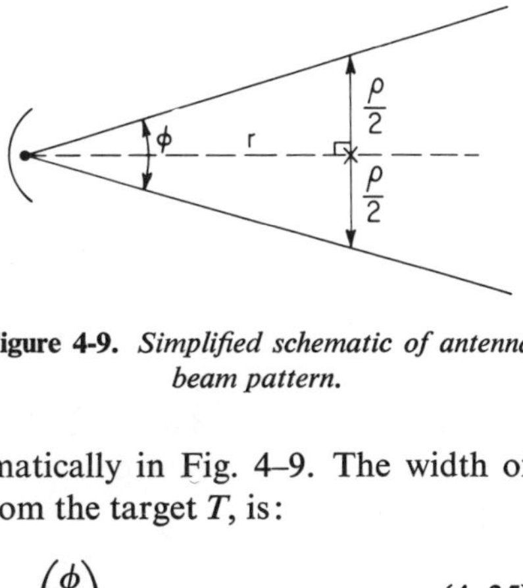

Figure 4-9. *Simplified schematic of antenna beam pattern.*

The antenna pattern is shown diagrammatically in Fig. 4–9. The width of the beam for intercept I_1, a distance r_1 from the target T, is:

$$\rho_1 = 2r_1 \tan\left(\frac{\phi}{2}\right) \quad (4\text{--}35)$$

and for I_2 is:

$$\rho_2 = 2r_2 \tan\left(\frac{\phi}{2}\right). \quad (4\text{--}36)$$

These two figures overlap in the target area. Figure 4–10 shows an enlargement of the detail. If it is allowed that the opposite sides of this figure are parallel over the range of intersection (which is not exact but is a reasonable approximation to simplify the geometry of the figure), the intersected area approaches a parallelogram. ρ_1 and ρ_2 are now perpendicular distances between opposite sides of the figure, as shown below.

The enclosed area of this figure is:

$$A = \frac{\rho_1 \rho_2}{\sin \alpha}. \tag{4-37}$$

Writing Eq. 4–37 in terms of the parameters of Fig. 4–8.

$$A = \frac{\rho_1 \rho_2}{\sin \alpha} = \frac{4 r_1 r_2 (\tan \phi/2)^2}{\sin (\theta_1 + \theta_2)}, \tag{4-38}$$

where

$$\alpha = 180° - (\theta_1 + \theta_2).$$

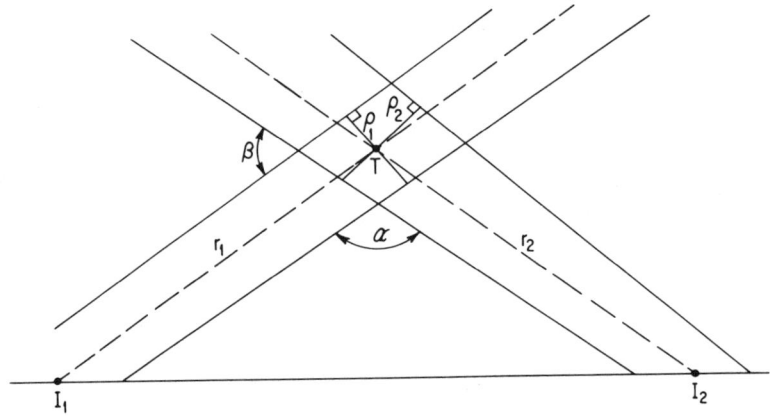

Figure 4-10. *Enlargement of intersecting area produced by two antenna patterns. Opposite sides of the figure are assumed parallel between the range of intersection.*

The first intercept (I_1) taken establishes θ_1 and r_1 as constants. However, the magnitude of r_1 is of course still unknown. r_2 is a function of θ_2 and can be written:

$$r_2 = \frac{R}{\sin \theta_2}, \tag{4-39}$$

$$\therefore A = \frac{4 r_1 R (\tan \phi/2)^2}{\sin (\theta_1 + \theta_2) \sin \theta_2}, \tag{4-40}$$

which gives the "area of uncertainty" as a function of one variable, θ_2 with respect to the parameters of $I_1(r_1, \theta_1)$.

To establish the minimum area, Eq. 4–40 is differentiated as follows:

$$\frac{dA}{d\theta_2} = \frac{-4 r_1 R (\tan \phi/2)^2 (\sin \theta_1 \cos 2\theta_2 + 2 \cos \theta_1 \sin \theta_2 \cos \theta_2)}{[\sin(\theta_1 + \theta_2) \sin \theta_2]^2}. \tag{4-41}$$

Setting the right-hand side of Eq. 4–41 equal to zero and reducing:

$$\sin \theta_1 \cos 2\theta_2 + 2 \cos \theta_1 \sin \theta_2 \cos \theta_2 = 0,$$

or

$$\sin \theta_1 \cos 2\theta_2 + \cos \theta_1 \sin 2\theta_2 = 0$$

$$\frac{\sin \theta_1}{\cos \theta_1} = -\left(\frac{\sin 2\theta_2}{\cos 2\theta_2}\right) \quad (4\text{–}42)$$

$$\tan \theta_1 = -\tan 2\theta_2,$$

but

$$\tan (n\pi - \theta_1) = -\tan \theta_1 = \tan 2\theta_2,$$

where $n = 0, \pm 1, \pm 2, \pm 3, \ldots,$

thus

$$n\pi - \theta_1 = 2\theta_2,$$

or

$$\theta_2 = \frac{n\pi - \theta_1}{2}.$$

To determine the value n observe that:

$$0 \leq \theta_1 + \theta_2 \leq \pi; \quad 0 \leq \theta_1 \leq \pi;$$
$$0 \leq \theta_2 \leq \pi,$$

$$\therefore \quad \frac{\theta_1}{2} + \frac{n\pi}{2} \leq \pi,$$

where $0 < n < 2$; hence $n = 1$

$$\therefore \quad \theta_2 = \left[\frac{\pi - \theta_1}{2}\right].$$

(4–42(A))

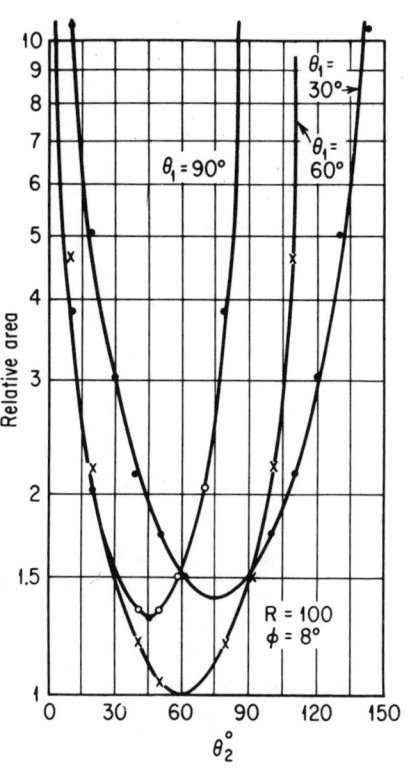

Figure 4-11.

Equation 4–42(A) defines the optimum angle for the I_2 bearing to produce the minimum intercepted area. For example, if the first intercept occurred at $\theta_1 = 30$ degrees the minimum area of uncertainty will result if the I_2 intercept is taken at $\theta_2 = 75$ degrees. Figure 4–11 shows a plot of the variation in area as a function of θ_2 for three values of θ_1. Note that for $\theta_1 = 30$ degrees, if θ_2 is also taken at 30 degrees, instead of 75 degrees, the area of uncertainty is three times the minimum. The plots shown in Fig. 4–11 represent a family of curves for variations of θ_1. Each curve has a slightly different minimum. It is now desirable to establish which θ_1 when combined with its optimum θ_2 will define the minimum areas of all possible combinations.

This result can be obtained by rewriting Eq. 4–40 and substituting in Eq. 4–42(A) as follows:

$$A = \frac{4r_1 R \left(\tan \frac{\phi}{2}\right)^2}{\sin\left(\theta_1 + \frac{\pi - \theta_1}{2}\right) \sin\left(\frac{\pi - \theta_1}{2}\right)}. \tag{4-43}$$

Simplifying:
$$A = \frac{4r_1 R (\tan \phi/2)^2}{\sin (\pi/2 + \theta_1/2) \sin (\pi/2 - \theta_1/2)}$$

which reduces to:
$$A = \frac{8r_1 R (\tan \phi/2)^2}{1 + \cos \theta_1}. \tag{4-43(A)}$$

However, r_1 is no longer a constant when θ_1 is permitted to vary. From Fig. 4–8,

$$r_1 = \frac{R}{\sin \theta_1}. \tag{4-44}$$

Substituting this relationship in Eq. 4–43(A) yields:

$$A = \frac{8R^2 (\tan \phi/2)^2}{(1 + \cos \theta_1) \sin \theta_1}. \tag{4-45}$$

To determine the minimum area as a function of θ_1 the derivative is taken:

$$\therefore \quad \frac{dA}{d\theta_1} = \frac{8R^2 (\tan \phi/2)^2 [\cos \theta_1 + \cos 2\theta_1]}{[(\cos \theta_1 + 1) \sin \theta_1]^2}. \tag{4-46}$$

Setting Eq. 4–46 equal to zero and simplifying gives:

$$\cos \theta_1 = -\cos 2\theta_1. \tag{4-47}$$

This relationship is satisfied for $\theta_1 = 60$ degrees. With θ_1 set at 60 degrees, the minimum area results with θ_2 taken at 60 degrees as indicated by Eq. 4–42 or Fig. 4–11. These conditions will provide the minimum area of uncertainty for two intercepts, with a given antenna, when taken from an aircraft flown in relation to the target emitter as shown in Fig. 4–8.

If optimum intercept angles are used ($\theta_1 = 60$ degrees, and $\theta_2 = 60$ degrees) the intercept area as a function of perpendicular distance, R, to target and antenna beam width, ϕ, can easily be established by inserting relationship 4–44 into Eq. 4–40 and substituting $\theta_1 = \theta_2 = 60$ degrees as follows:

$$A = \frac{4R^2 (\tan \phi/2)^2}{\sin (\theta_1 + \theta_2) \sin \theta_2 \sin \theta_1} = \frac{4R^2 (\tan \phi/2)^2}{\sin 120° \sin 60° \sin 60°}, \tag{4-48}$$

$$\therefore \quad A_{\min} = \frac{4R^2 (\tan \phi/2)^2}{(0.866)^3} = 6.2R^2 (\tan \phi/2)^2. \tag{4-49}$$

An aircraft carrying antennas with 8-degree beam width, intercepting an emitter 100 miles from the closest point on its flight path (and flying a straight course) would locate the target to within: $A = 6.2(100)^2(\tan 4°)^2 = 304$ miles². In Fig. 3–11 this figure will represent the minimum relative area of 1. If θ_1 and θ_2 were each 30 degrees the relative area is shown as three times as great or, $3 \times 304 = 912$ miles².

This problem has special interest when it is understood that in some high-speed reconnaissance aircraft it is required that the antennas be flush mounted in a fixed position. When two antennas are used as shown in Fig. 4–12, a knowledge of the mounting angles in relation to the flight line of the

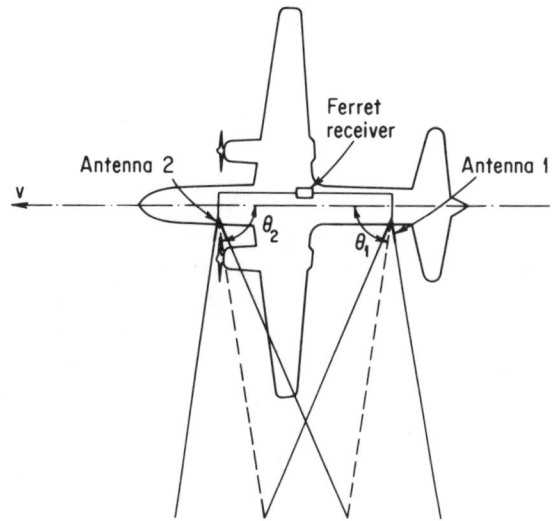

Figure 4-12. *Reconnaissance aircraft carrying two flush-mounted surveillance antennas fixed at angles θ_1 and θ_2 relative to flight path.*

aircraft can establish the minimum area of uncertainty that can result from such paired intercepts.

A major difference between the results of the horizontal scan mode and the overflight mode lies in the fact that a single intercept in the overflight mode produces a finite area of uncertainty, when a minimum of two intercepts is needed in the horizontal mode.

INTERCEPT CORRELATION

The discussion up to this point has concerned itself with the geometries and probabilities associated with the interception of an emitter. The signal radiated by the emitter is of course made up of many measurable characteristics. Heretofore, the statement that an intercept was accomplished implied that the desired signal parameter was recorded. In general, it is advantageous to record more than a single parameter and thereby increase the information gained from each intercept. The objective of the mission will affect which, and how many, of the signal parameters should be recorded.

The list of characteristics associated with an emitter that might be recorded could include:

1. Carrier frequency.
2. Pulse-repetition frequency (PRF).
3. Pulse width (PW).
4. Antenna scan rate.
5. Antenna scan pattern.
6. Antenna side-lobe structure.
7. Message contents.
   (a) AM
   (b) FM
   (c) Digital.

Practical considerations in the design of equipment, however, will not allow the recording of all of these characteristics simultaneously. Further, it is not generally necessary to record so many parameters to obtain the desired information. The question then arises as to which parameters should be chosen to provide the desired information with the highest confidence factor.

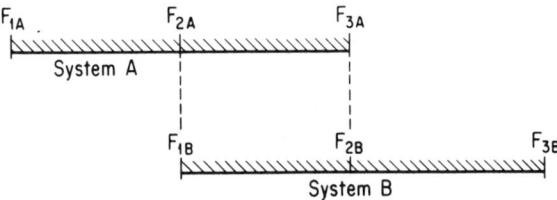

Figure 4-13. *Partial co-frequency operation of System* A *and System* B.

As an example, assume that the enemy is using two types of systems; System A and System B. The number of system A's in use must be identified; there is no interest in System B operation. Also, the only characteristic to be measured will be the frequency of emission. Let System A operate anywhere between a frequency of F_{1A} and F_{3A} and let System B operate anywhere between a frequency of F_{1B} and F_{3B}. This situation is shown in Fig. 4-13, where $F_{2A} = F_{1B}$ and $F_{3A} = F_{2B}$. It can be seen that the overlapping use of frequencies does not permit the positive identification of System A operation based on this parameter alone. In fact, the probability that a System A signal will be correctly identified is

$$P_{ID} = \frac{F_{2A} - F_{1A}}{F_{3A} - F_{1A}} + 0.5\left(\frac{F_{3A} - F_{2A}}{F_{3A} - F_{1A}}\right). \quad (4\text{-}50)$$

Should the two systems use exactly the same frequencies (Fig. 4–14), the probability of correct identification would be:

$$P_{ID} = \frac{F_{2A} - F_{1A}}{(F_{2A} - F_{1A}) + (F_{2B} - F_{1B})} = 0.5. \quad (4\text{-}51)$$

In this case, if 100 systems of each type were in use, the count of System A might vary anywhere between 0 systems and 200 systems. It is impossible to identify System A operation directly under these conditions.

In this simple example, frequency is not a good parameter choice if system identification is the objective. However, if a measurement of signal density is the objective, the frequency is a very proper parameter to record.

The mission objective will dictate what parameters to measure, and to what degree of accuracy they must be measured. When the requirement for more precise information indicates that two (or more) parameters should be measured, consideration of the correlation factor between respective parameters should be studied.

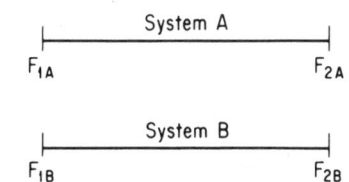

Figure 4-14. *Total co-frequency operation of System* A *and System* B

A simple example will illustrate this point. Let there be in fact 50 System A's and 100 System B's in operation and say that 150 intercepts (one of each) are taken. For the first set of intercepts let the parameters recorded be frequency and pulse width. From these measurements assume that 90 intercepts are identified as System A and 60 intercepts are identified as System B. In the second case, let the parameters recorded be frequency and pulse-repetition frequency. From these measurements assume that 55 intercepts are identified as System A and 95 intercepts identified as System B. Obviously, the second case has produced results much closer to the real situation (50 System A's, 100 System B's). Therefore it is concluded that frequency/PRF would have a much higher correlation factor than frequency/PW. Indeed, studies have been carried out showing that a higher correlation factor will result from a comparison of one pair of parameters with another pair when both are applied to various forms of emitter identification. It is a requirement placed on both the design and the application of a ferret system that it correctly select the signal characteristics to be recorded during a reconnaissance program.

GENERALIZED "SENSOR" CHARACTERISTICS

It is now possible to consider some of the requirements and restraints placed on a reconnaissance program by the ferret receivers. Even when all the external conditions for an intercept are met, a great many requirements must be met by the ferret receiver. These requirements are not always mutually compatible and force compromises in the practical design. Because of the limitations in performance of a single receiver it is quite common to use a multiplicity of units on a reconnaissance mission. Each receiver in the complex can then be utilized with optimal efficiency in the accomplishment of its particular function.

This discussion, however, will consider the limits of performance that might be reasonably expected from a single receiver.

Perhaps the most important performance requirement to be considered in connection with a reconnaissance receiver is the total Band Width (BW) it is capable of placing under surveillance. Theoretically, it would be desirable to have one receiver capable of covering the entire RF spectrum. At the same time, of course, the receiver could also be restricted to any portion of the total spectrum if so desired. In practice, unfortunately, receivers are generally restricted by circuitry limitations to octave coverage. Nevertheless, the trend is to build receivers with as large a BW coverage as possible to conserve weight and space.

Along with the large BW coverage requirement, a narrow Detection Band Width (DBW) is desired. The detection band width must be able to pass the information being carried in the signals received and this minimum will be designated $\overline{DBW}$. Therefore, DBW must be equal to or greater than $\overline{DBW}$. However, it is the DBW that determines the frequency resolution obtainable. The requirements of the reconnaissance program dictate what frequency resolution is necessary. In general, a resolution of 5% is quite adequate (there are indeed exceptions to this number). This parameter can be simply defined as:

$$\% \text{ detection resolution} = \frac{IF \text{ band width}}{RF \text{ signal frequency}} \times 100. \qquad (4-52)$$

From this definition a signal recorded by a receiver, whose dial calibration is believed accurate and reads 3000 mcs with a 30 mcs final IF, produces a detection resolution of:

$$\frac{30 \times 10^6}{3 \times 10^9} \times 100 = 1\%.$$

The signal is therefore interpreted as being at 3000 mcs $\pm$ 15 mcs. The purpose of specifying a signal frequency within a reasonable tolerance lies in the fact that it allows relocation of the carrier frequency at a later time for the purpose of intentional jamming.

An additional factor strongly affecting the receiver operation is its overall sensitivity. It is reasonable to expect that some minimum sensitivity will be incorporated into the receiver design. A great deal has been written about receiver sensitivity considerations and the subject will not be treated here at any length. However, one method of expressing the receiver sensitivity is as equivalent noise power referred to the input terminals. This noise power may be written as:

$$NF \cdot k \cdot T \cdot \Delta F \text{ watts} \qquad (4\text{–}53)$$

where NF is the receiver noise figure,
$\quad k$ is Boltzman's constant (1.38×10^{-23} Joules/degree Kelvin),
$\quad T$ is absolute temperature (°K),
$\quad \Delta F$ is IF band width (cps).

Noise power in turn is usually expressed in decibels above a milliwatt (dbm) with respect to the equivalent noise of some reference source (1 mw in a 50 ohm resistor). Therefore, the receiver sensitivity in dbm is:

$$\text{dbm} = 10 \log_{10} \frac{NF \cdot k \cdot T \cdot \Delta f}{10^{-3}}. \qquad (4\text{–}54)$$

This number is generally negative; typical receivers have sensitivities in the range -90 dbm to -125 dbm, depending on the frequency being considered. Again, there are exceptions to this range and receivers have been built with sensitivities in the order of -150 dbm.

The three characteristics discussed so far are of concern in the design of receivers in general. The next characteristic to be considered applies specifically to the ferret receiver. This characteristic will be referred to as the Dwell Time (DT) of the system. Because of the logic circuitry, recording limitations, and methods of electronic scanning (if used) a ferret receiver is not always ready to accept every signal that falls within its domain. In other words, it is dwelling on a different information source. This, of course, is a way of stating that the information-handling rate of the receiver is limited. It is desirable to have the DT approach zero, thereby indicating that the receiver is always ready to accept a new signal. The difference in acceptance time between two signals Δf cycles apart is a measure of the dwell time (DT).

Chart 4–1 provides a summary of the four parameters discussed.

Chart 4–1

	Parameter	Symbol	Characteristic affected	Desired general range
1.	Total band width	BW	Spectrum observed	Large
2.	Detection band width	DBW	Resolution provided	$\approx \overline{DBW}$
3.	Sensitivity	dbm	Signals detectable	High
4.	Dwell time	DT	Acceptance time required per signal	Low

It is possible to show a functional relationship between these factors by the introduction of a figure of merit, K.

$$K = f(BW, DT, dbm, DBW/\overline{DBW}) \qquad (4\text{-}55)$$

where, as before, $\overline{DBW}$ is the required information band width.

To apply relationship (4-55) to a quantitative problem, an understanding of the relative importance of the parameters to the specific mission objective is needed. Equation 4-55 can then be written with the appropriate weighting factors which will permit quantitative comparisons between receiver types.

RECEIVER MODES

One general class of receivers used for reconnaissance is referred to as the "wide-open" or video receiver. As the name implies, this receiver provides a wide band-pass by means of broadly tuned circuits. The wide band-pass allows the information contained in the high-frequency components of the signal (pulses) to be faithfully reproduced during the amplification, detection, and recording operations. For example, when the rise and fall time of a radar pulse is of interest, this receiver provides an excellent method of observing these characteristics. As the total BW of this system is increased, the signal amplification available tends to decrease as a result of the gain-band width limitation imposed by a given tube. Since no frequency conversion is used in this receiver, all amplification is provided at the signal frequency. There are, of course, practical limits to the amount of gain that can be provided before a tendency toward oscillation occurs. As a result of this limitation, the sensitivity of this system is relatively low. However, a decided advantage of this type of receiver is its fast signal-handling capability. Having no scan or search circuitry, and generally a minimum (if any) logic circuitry, any signal appearing within the range of its BW is immediately available for recording. Unfortunately, in most video receivers the frequency resolution is no better than the total input band width.

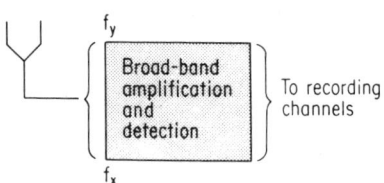

Figure 4-15. *Simplified block diagram of video receiver.*

To overcome the handicaps of low sensitivity and poor frequency resolution of the video receiver, a technique that might be referred to as a paralleled filter system can be used. In truth, this method is not too different from taking the desired frequency coverage, f_x to f_y shown in Fig. 4-15, and dividing it up into smaller bands. Then each smaller band may be covered by a narrower band width video receiver, increasing both sensitivity and frequency resolution. Although the paralleled filter receiver does essentially that, it makes use of certain circuitry in common, thus reducing the redundancy of equipment.

Since this receiver is rather complex, a numerical example together with the block diagram shown in Fig. 4–16 will serve to illustrate the general principle. Assume that the front end of this receiver is comprised of two broad channels: the channel C_1 covering 10,000 mcs to 11,000 mcs, and the second channel C_2, covering 11,000 mcs to 12,000 mcs. Each of these channels is 1000 mcs wide. A signal passing through either channel (C_1 or C_2) is now converted to a frequency between 1000 mcs and 2000 mcs. This frequency range is divided into three channels: C_3, C_4, and C_5. The corresponding frequencies are 1000–1333 mcs, 1333–1666 mcs, and 1666–2000 mcs respectively. Each of these channels is 333 mcs wide. A signal passing through

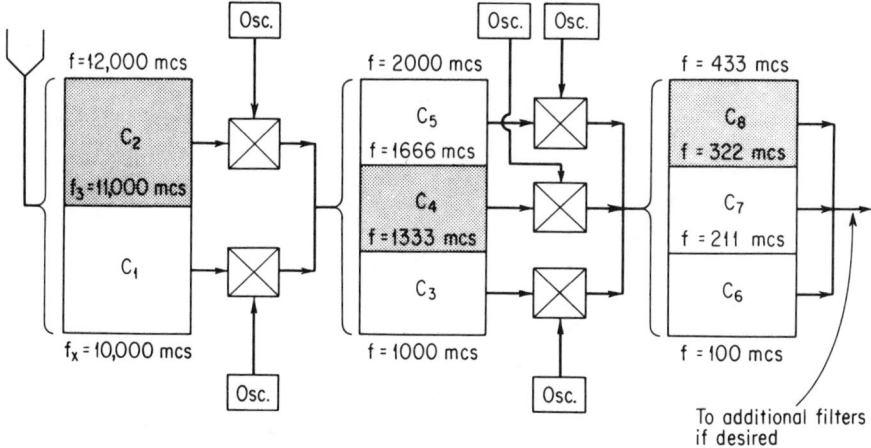

Figure 4-16. *Simplified block diagram of paralleled filter receiver.*

any of these channels (C_3, C_4, or C_5) is in turn converted to a frequency between 100 mcs and 433 mcs. This frequency range is also divided into channels C_6, C_7, and C_8. The corresponding frequencies here are 100–211 mcs, 211–322 mcs, and 322–433 mcs respectively. Each of these channels is 111 mcs wide. This process may be carried on until the final channel width is as narrow as desired.

When an incoming signal passes through a channel its passing is recorded by logic circuitry. Assume now that a signal entered the receiver and passes through channels C_2, C_4, and C_5 as shown by the shaded area in Fig. 4–16. The fact that it passed through channel C_2 indicates that the signal frequency lies somewhere between 11,000 mcs and 12,000 mcs. This range might be about equivalent to the frequency resolution a video receiver would produce. However, here the signal also passed through channel C_4. Its frequency is therefore limited to between 11,333 mcs and 11,666 mcs. Finally, it passed through channel C_8, defining its frequency as being in the upper 111 mcs of the range covered by channel C_4. Therefore, the signal frequency must lie between 11,555 mcs and 11,666 mcs. This system has provided a frequency resolution of approximately 1% and may be carried further if desired. This is in keeping with the results suggested by Eq. 4–52.

This system is really nothing more than a series of channelized filters stacked together, provided with a record as to which filter is being used by the incoming signal. A closely similar approach is used with some common types of audio-frequency analyzers. However, this method does provide improved sensitivity and frequency resolution over the straight video receiver. Unfortunately, in an active electronic environment the dwell time may be rather long, since a signal passing through a particular channel must inhibit the channel (and therefore that segment of the spectrum) until released.

A third type of system to consider is the scanning receiver. In its simplest form it can consist of a superheterodyne receiver being manually tuned across the band of interest by an operator. Indeed, during the early part of World

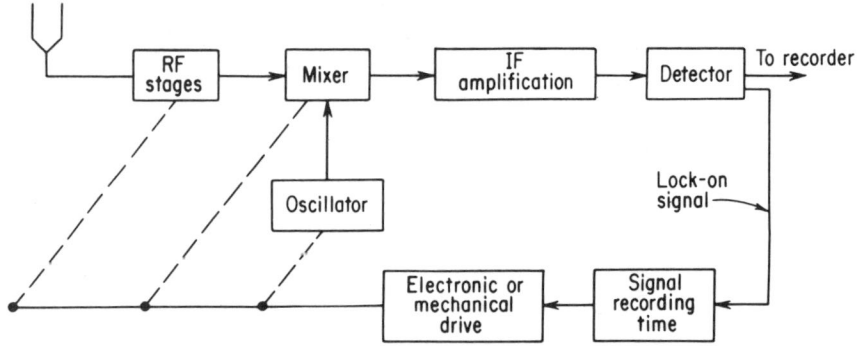

Figure 4-17. *Scanning superheterodyne receiver.*

War II operators manually searched the bands and recorded all signals of interest with pen and paper. Basically, all scanning receivers accomplish their function by tuning across the band. The tuning method, however, generally consists of some complex electronic or mechanical system that is capable of recognizing when a signal is received and that can stop the scanning cycle for some fixed length of time.

After the signal has been recorded the scanning is started again. In its simplest form a block diagram (Fig. 4–17) for this receiver will appear similar to a conventional superheterodyne receiver with the addition of some form of scanning drive system. It can be seen from Fig. 4–17 that the *RF*, mixer, and oscillator stages are all tuned together, providing the tracking alignment generally necessary for good receiver sensitivity. As a result of this mode of operation the receiver provides higher sensitivity than either of the types discussed earlier.

The signal-handling rate of this method is somewhat limited in that it is in effect moving a "window" across the *RF* spectrum to be scanned. A representation of this is shown in Fig. 4–18. Since the rate at which this scanning can be accomplished is limited, it is possible for a signal to be present at a point within the band while the receiver is tuned elsewhere. This condition is also shown in Fig. 4–18, when the receiver is at a frequency f' and a signal at f''. It is possible that by the time the receiver reaches frequency f'' the signal will no longer be present. To be sure, the rate of scan can be made very fast,* but this fast sweep in turn causes a loss in both sensitivity and frequency resolution. This loss results from the fact that a tuned circuit moved across a signal at too high a rate in relation to its band width produces a ringing, or shock-excited effect, which tends to obscure the true characteristics of the desired signal. This problem can be offset by broadening the *IF* band width. However, the noise is increased in turn

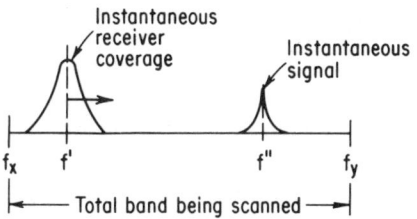

Figure 4-18. *Receiver bandpass at f' being swept toward signal at f''.*

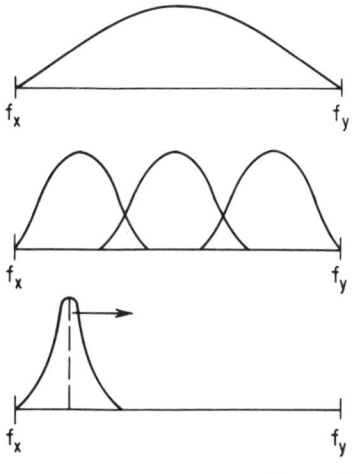

(A) Single broad-band tunned circuit (video)

(B) Many tunner circuits placed side by side (paralleled filter)

(C) Sharply tunned circuit moved across the band (scanning receiver)

Figure 4-19.

* An analysis of the scan rate of this type of receiver is exceedingly complex. An excellent discussion and mathematical treatment of the problem is given in the "Handbook of Spectrum Analysis Techniques" by Polard Electronics Corporation, 43–20 34th Street, Long Island City 1, N.Y. It is shown here that the loss in sensitivity as a function of sweep rate is given by:

$$a_s = \left[1 + 0.195\left(\frac{F}{TB^2}\right)^2\right]^{-1/4}$$

where a_s is the loss in db relative to zero sweep rate,
  F is the sweep width (dispersion) in cycles per second,
  T is the sweep time interval in seconds,
and therefore F/T is the sweep rate in cycles per sec per sec.

(Eq. 4–54), thereby reducing the sensitivity and also degrading the detection resolution (Eq. 4–52).

The decision as to which way the design should go in connection with these problems must be based on the consideration of the relative importance of the respective factors, as discussed in connection with Eq. 4–55.

We have, then, covered briefly three types of receivers that could be used successfully for reconnaissance work. In all cases they can be compared with the tuned-circuit problem for covering a given band of frequencies. Figure 4–19 gives a representation of this approach.

In this chapter we have studied some of the problems of acquiring information in connection with the concept of electronic warfare. In the following chapters we shall discuss the problems of ECM and ECCM and how to achieve optimal effectiveness in their use in the light of information gained from reconnaissance programs and other sources.

5

RADAR CONSIDERATIONS

THE APPROACH TO RADAR COUNTERMEASURES

Historically, electronic warfare has been principally concerned with techniques for seeking out the enemy's targets in either normal or countermeasure environments, or in preventing the enemy from detecting our targets. Electronic areas other than target detection, such as communications or navigation, have received less emphasis. Radar and radar countermeasures were the sources of much of the electronic-warfare effort during World War II. Some idea of the part that radar played is indicated by the fact that approximately $3 billion worth of radar equipment was developed during the war by the United States alone. The production rate at the end of the war was over $100 million worth per month. It is to be expected that any technique as valuable to warfare as radar should provoke a strong effort to nullify its performance. The fact that radars are unusually vulnerable to countermeasures makes their role in electronic warfare all the more important.

Any consideration of the methods by means of which countermeasures can be used against radars requires an understanding of the various types of radar systems and their principles of operation. Once the normal method of operation is known, the techniques for countering this normal operation are often obvious. However, there are a variety of different classes of radars, such as pulse, Continuous Wave- (CW) and Pulse-Doppler (PD), and each type makes use of a variety of different techniques that are vulnerable to varying degrees. Generally, an ECM system is designed to operate against a particular type of radar and to counter certain techniques only. Then, as new types of radars are developed, new ECM systems must be devised to handle them. In spite of this continuous "leap-frogging" of radar and counter-

82 RADAR CONSIDERATIONS

radar systems, it is still possible to develop certain fundamental principles that will apply to nearly all radar and ECM situations. In this chapter we shall develop some of these radar principles. We shall also discuss the basic types of radar systems in enough detail to give an understanding of how they operate and how they can be countered.

PULSE RADARS

The first radar systems were pulse radars, as were nearly all the radars used during World War II. The majority of radars presently in operation use these same pulse techniques with no essential change in principle. A pulse of radio energy is transmitted and a receiving system detects any echoes from targets. A selective antenna determines the direction to the target while the round-trip propagation time of the pulse gives the range. Since the velocity of light is very close to 1000 feet/μsec, the range is:

$$R \text{ (nautical miles)} = \tfrac{1}{12}t \text{ }(\mu\text{sec}). \tag{5-1}$$

Almost all pulse radars wait long enough for any echoes to return from the maximum range of interest before transmitting the next pulse, so the above relationship also sets the radar pulse-repetition rate. For example, 1000 pulses/sec have 1000 microseconds between pulses, providing a maximum unambiguous range of approximately 80 miles.

Radars operate at short radio wavelengths where it is easier to focus their energy into narrow beams. Also, the reflectivity of targets decreases rapidly when their dimensions become much smaller than a wavelength. The principal radar bands have been given letter designations. Their frequency ranges (in megacycles) are:

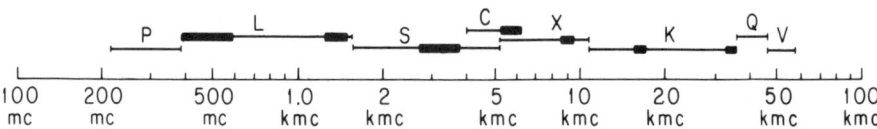

Figure 5-1. *Radar bands. The bars indicate the most commonly used frequencies.*

Although these ranges are rather approximate, radars are often classified according to band. Many other non-radar communications systems also operate within most portions of the bands.

PULSE-RADAR OPERATION

A schematic block diagram for a typical pulse radar is shown in Fig. 5-2. Radar energy is generated by a magnetron oscillator powered with high-voltage pulses supplied by the modulator. The modulator takes in energy

PULSE-RADAR OPERATION 83

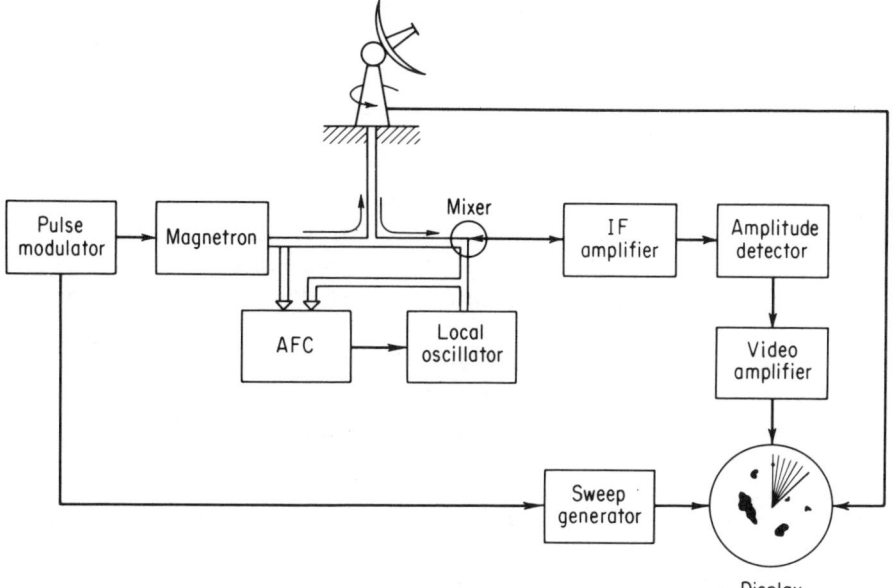

Figure 5-2. *Non-coherent pulse radar.*

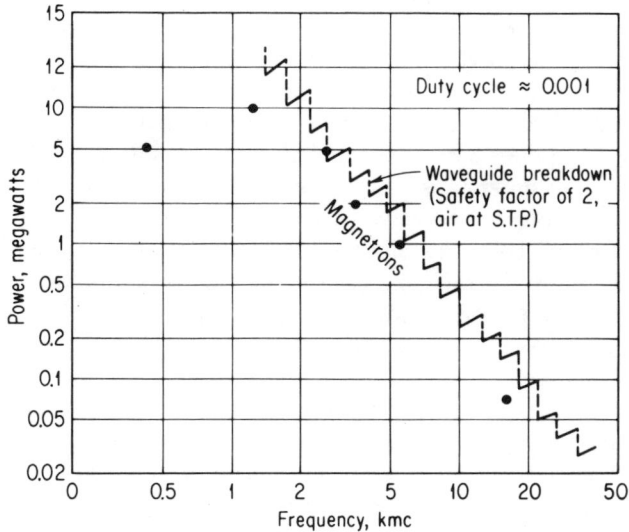

Figure 5-3. *Typical peak power output for high-power radars.*

continuously, stores it in capacitors or inductors, and releases it in short bursts. For most pulse radars the duty factor (ratio of average to peak powers) is of the order of 0.001. All pulse radars do not use magnetrons. Klystrons and conventional high-power vacuum tubes are used at lower frequencies and magnetrons are used at the higher frequencies. The dividing line is roughly 1000 mc. However, the magnetron is the most common pulse-radar power generator.

The power is transmitted to the antenna by a coaxial line at the lower radar frequencies or by a waveguide at the higher frequencies. Arcing in the waveguide limits the peak power that can be transmitted, as shown in Fig. 5-3. Pressurizing the gas in the line will increase the power-handling capability of the system.

The angular resolution of the radar is obtained with a directional antenna that focuses the radar energy into a narrow beam. As an example, at 10,000 mcs an antenna 6 ft in diameter has a power gain of 20,000 times and a 1-degree beam width. At either side of the main lobe the antenna gain is considerably reduced, but by no means to zero. The main-lobe and side-lobe levels of antennas are important ECM considerations, since they determine the extent to which a radar antenna can reject ECM coming from other than the target direction. Antenna characteristics are discussed in Chapter 6.

During the search mode, the antenna systematically scans through the solid angle in which targets are expected. Since it is usually assumed that targets are equally likely to be in any portion of this solid angle, the antenna spends no more time looking in one direction than in any other. However, the solid angle searched can usually be changed, corresponding to what is known about likely target locations. The antenna is used for both transmitting and receiving, so it must not rotate the beam off the target before the echo returns. In practice, antenna speeds are designed so that many echoes can be received during each sweep past a target. A radar having a 1-degree antenna, rotating at 10 rpm, and a pulse-repetition frequency of 600 pulses/sec would receive 10 pulses between beam half-power points.

Since the same antenna is used for receiving as well as for transmitting, it is necessary to protect the receiver from high overloading signals during the time the pulses are being transmitted. This protection is accomplished by a series of *T-R* (transmit-receive) duplexing devices: gas switches that operate by ionizing during transmission, thus protecting the receiver from burnout. The details of duplexing will not be covered here.*

When a radar echo is received, it is delivered to a microwave mixer, where it is heterodyned or "mixed" with a Local Oscillator (*LO*) signal to generate an Intermediate Frequency (*IF*) signal. The process is one of converting the radar signal to a lower frequency that can be more easily amplified. The mixer is a crystal diode inserted in the waveguide or coaxial line. The sum

---

* MIT Radiation Laboratory Series, *Microwave Duplexers*, Vol. 14, McGraw-Hill Book Company, Inc., 1948.

of the local oscillator signal plus the echo signal is fed to the mixer. The mixing operation of the diode can be visualized as squaring the sum of the two signals, as:

$$(e_S + e_{LO})^2 = e_S^2 + e_{LO}^2 + 2e_S e_{LO}. \tag{5-2}$$

The cross-product term, $2e_S e_{LO}$, represents the multiplication of the echo by the local oscillator signal. Multiplying two signals of different frequencies results in the generation of "sum" and "difference" frequencies:

$$2(\cos \omega_S t) \times (\cos \omega_{LO} t) = \cos (\omega_S + \omega_{LO})t + \cos (\omega_S - \omega_{LO})t. \tag{5-3}$$

In practice, the sum frequency is discarded, and the difference frequency becomes what is called the "*IF*." Although this explanation of mixing is simplified, it does illustrate how the difference frequency, or *IF*, arises. Mixing is best regarded as a shifting of the input signal frequency by the amount of the local oscillator frequency to the intermediate frequency. Except for its new frequency (and some loss in power), the signal is otherwise unchanged. An excellent mixer can be defined as a device that multiplies the two input signals, and this technique will be used in later sections of this chapter.

The local oscillator is usually a reflex klystron operating at a frequency that differs from the radar frequency by the *IF*, which is commonly about 30 or 60 megacycles. To keep the klystron from drifting off frequency, the magnetron and local oscillator signals are fed to an automatic frequency-control system, which mixes them to generate an *IF* signal, which is amplified and applied to a frequency discriminator that senses any variation of this *IF* from the desired *IF*. The error signal so developed is used to retune the klystron to the proper *LO* frequency.

The *IF* amplifier increases the echo signal to a level at which it can be detected, which usually requires a voltage gain of about 1,000,000 times. This gain will also amplify any noise present in the antenna, mixer, or first stages of the *IF* amplifier. The signal must be stronger than the noise if it is to be detected, as discussed in Chapter 3. Thus, noise places a fundamental limitation upon the sensitivity of the radar receiver. Even if the receiver and mixer do not generate noise, the effective resistance of the antenna will provide a source of thermal noise. The amount of this noise power that is coupled into a receiver from the antenna that will pass through the *IF* amplifier pass band is $kT\Delta f$, where k is Boltzmann's constant, T is temperature in degrees Kelvin, and Δf is the band width in cycles/sec (kT is about 4.2×10^{-21} w-sec). This ideal situation is never met in practice, since the mixer and *IF* amplifier always generate some noise, so the noise power intercepted by the *IF* amplifier is greater than $kT\Delta f$ by a factor defined as the "noise figure," *NF*. Thus it is assumed that the receiver is "ideal" and that the noise from the antenna is:

$$P_n = kT\Delta f \overline{NF}. \tag{5-4}$$

This is the same result arrived at in Chapter 4 under the discussion of the reconnaissance receiver sensitivity (Eq. 4–53). The noise figure is a function of frequency, as is shown in Fig. 5–4.

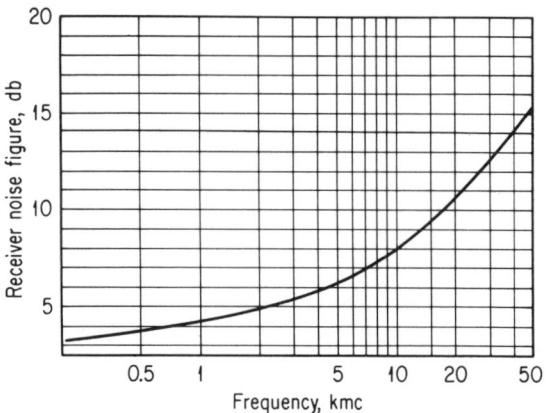

Figure 5-4. *Typical receiver noise figure.*

OPTIMUM *IF* AMPLIFIER BAND WIDTH

The *IF* amplifier must have a band width sufficiently wide to pass the narrow signal pulses, yet if it is made too wide it will also pass excessive noise. A rectangular radar pulse is shown in Fig. 5–5. It is a sinusoidal voltage $E_c \cos(2\pi f_c t)$ chopped into pulses of duration τ_o and at a pulse rate of f_r pulses/sec. The resulting frequency spectrum is also shown. It consists of a series of spectral components spaced apart in frequency by the repetition rate, and centered about the carrier frequency, f_c. Their amplitudes vary as $(\sin x)/x$, gradually diminishing either side of f_c. The amplitude of f_c is $f_r \tau_o E_c$ volts peak. Each component by itself is a single distinct sinusoidal signal, not merely a mathematical fiction. A narrow-band receiver tuned to any one component would find it a continuous wave, unmodulated. By definition, if all the spectral components were added together they would produce the original short pulses. Some idea of just how this result comes about can be had by picturing each component as a vector rotating about the origin. The central component rotates f_c times per sec, the next one to the right rotates $(f_c + f_r)$, the next $(f_c + 2f_r)$, and so on. Thus, the relative phases between all

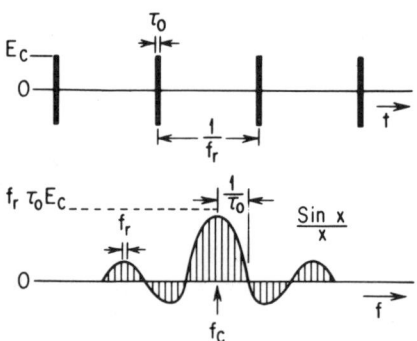

Figure 5-5. *Radar pulses and the resulting spectrum.*

the vectors are continuously changing. Once during every period $1/f_r$, the vectors align themselves for an interval τ_o so they are all in a straight line (the positive ones adding, the negative ones subtracting), which adds up to a voltage E_c. During the remaining portion of each period, however, the vectors spiral around in such a manner that their resultant is exactly zero.

An infinite number of components is needed to represent a perfectly rectangular pulse. Suppose that only the components within a rectangular frequency band Δf wide, centered about f_c, are selected and the rest rejected.* The resulting pulse amplitude (at the center of the pulse) can be obtained by summing these spectral components:

$$E_{pk} = \sum_{f=f_a}^{f_b} E_c \tau_o f_r \frac{\sin x}{x} = E_c \tau_o f_r \sum_{f=f_a}^{f_b} \frac{\sin x}{x}$$

where
$$x = \pi(f - f_c)\tau_o \qquad f_b - f_a = \Delta f \qquad (5\text{-}5)$$
$$f_b = f_c + \Delta f/2.$$

When the number of components to be summed is large, the summation can be approximated by an integral (noting that f_r is the spacing between components).

$$E_{pk} \simeq E_c \tau_o f_r \frac{1}{f_r} \int_{f_a}^{f_b} \frac{\sin x}{x} df$$
$$\simeq E_c \tau_o f_r \frac{2}{\pi \tau_o f_r} \int_0^{\pi \tau_o \Delta f/2} \frac{\sin x}{x} dx \qquad (5\text{-}6)$$
$$\simeq E_c \frac{2}{\pi} Si(\pi \tau_o \Delta f/2)$$

where
$$Si(u) = \int_0^u \frac{\sin x}{x} dx.\dagger$$

For example,

$E_{pk} = E_c$ if $\tau_o \Delta f = \infty$ Corresponding to an infinitely wide band width.

$E_{pck} = 0.99E$ if $\tau_o \Delta f = 1.2$.   For optimum band width

---

* The assumption of a rectangular band-pass *IF* amplifier is idealized. Constant gain and linear phase shift within the band-pass result in no distortion of the signal components *within the band*, only time delay. However, this ideal filter cannot be physically realized since its response occurs before the excitation. Practical filters often approach this rectangular response, and the ideal filter is often assumed since it greatly simplifies computations. A more optimal type of filter for radar signals is discussed in the section entitled "Effects of ECM on Radar."

† The Sine Integral $Si(u)$ is a common function, tabulated in most mathematical handbooks.

Figure 5–6 shows a plot of peak pulse voltage as a function of $\tau_o \Delta f$. Equation 5–4 shows that noise power is proportional to Δf. Since power is proportional to voltage squared, noise voltage increases as $\sqrt{\Delta f}$. This proportion is shown

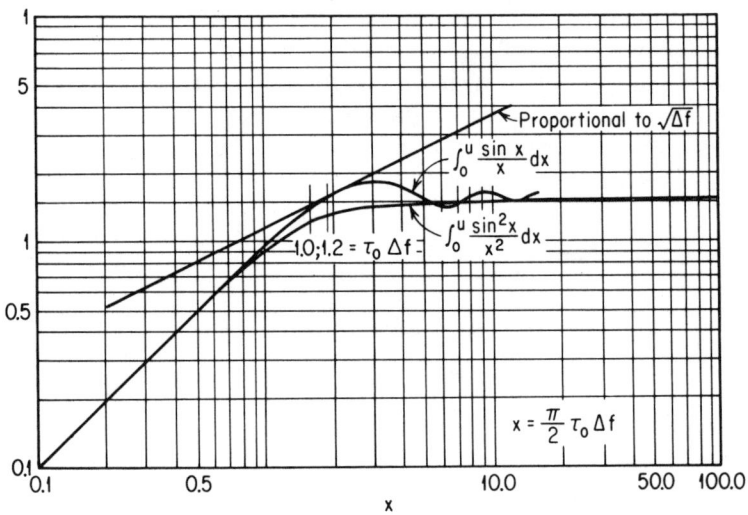

Figure 5-6. *The integral of $(\sin x)/x$ and $(\sin^2 x)/x^2$ as a function of x.*

as a line in Fig. 5–6. Note that either side of approximately $\tau_o \Delta f = 1.2$ the noise voltage increases faster than the pulse voltage. Thus, $\tau_o \Delta f$ should be about 1.2 for maximum signal-to-noise ratio out of the *IF* amplifier. When $\tau_o \Delta f = 0.5$ or 3, the ratio of signal-to-noise is reduced 3 db from the opti-

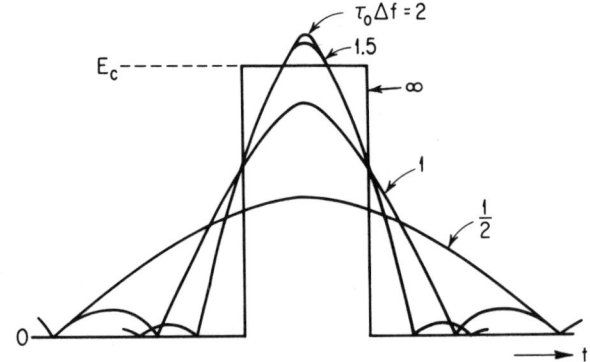

Figure 5-7. *Pulse distortion for various amplifier band widths.*

mum value, which has the same effect on detection range as reducing the transmitted power to one half.

The waveshapes out of the amplifier for several values of $\tau_o \Delta f$ are shown in Fig. 5–7. Notice that as the band width decreases and the pulses get smaller they also get wider, so that although the peak pulse voltage decreases proportionally with band width for $(\tau_o \Delta f) < 0.5$, the pulse power does not decrease proportionally with the square of band width, but directly with band width.

Although the ratio of *peak* signal to average noise decreases when the band width is narrowed, the ratio of *average* signal to average noise does not. This suggests that the *IF* band width can be made very narrow and the resulting S/N from the *IF* amplifier will still be just as satisfactory for target detection (of course the output will no longer consist of pulses). This conclusion will be verified in the section on pulse-Doppler radars. The average power contained in the spectral components within Δf can be computed by adding the powers of each spectral line:*

$$P_{av} = \frac{1}{2R} \sum_{f=f_a}^{f_b} \left(E_s \tau_o f_r \frac{\sin x}{x} \right)^2 \tag{5-7}$$

$$\simeq \frac{E_s^2}{2R} (\tau_o f_r)^2 \frac{2}{\pi \tau_o f_r} \int_0^{\pi \tau_o \Delta f/2} \frac{\sin^2 x}{x^2} dx$$

$$\therefore P_{av} \simeq \frac{E_s^2}{2R} \tau_o f_r \frac{2}{\pi} \left[S_i(\pi \tau_o \Delta f) - \frac{\sin^2(\pi/2)\tau_o \Delta f}{(\pi/2)\tau_o \Delta f} \right] \tag{5-8}$$

whence:
$$P_{av} = \frac{E_s^2}{2R} \tau_o f_r \quad \text{if} \quad \tau_o \Delta f = \infty$$

$$P_{av} = (0.84) \frac{E_s^2}{2R} \tau_o f_r \quad \text{if} \quad \tau_o \Delta f = 1.2.$$

The variation of average pulse power with $\tau_o \Delta f$ is plotted in Fig. 5–6.

PULSE DETECTION AND INTEGRATION

The *IF* amplifier is followed by an amplitude detector that demodulates or removes the pulse from its radio-frequency carrier. The commonest type of

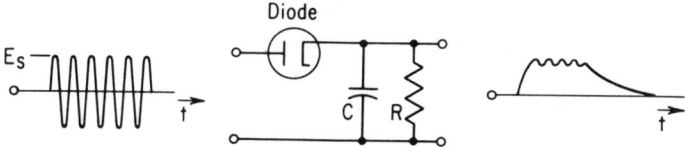

Figure 5-8. *Amplitude detection.*

amplitude detector, shown in Fig. 5–8, consists of a diode followed by a capacitor and resistor. The input pulse is a voltage consisting of many cycles (30 cycles if the *IF* is 30 megacycles and the pulse duration is one microsecond). During the positive half-cycle the diode conducts, charging the

* The signal is applied across an arbitrary resistance, R.

capacitor. After a few cycles the voltage on the capacitor has reached the peak voltage of the input pulse. The size of discharging resistance is chosen so that the capacitor discharges very little between cycles yet discharges rapidly when the pulse ends. The resulting output pulse has much the same shape as the envelope of the input pulse, but the frequency components in the output are different. The pulse spectrum of the detector input is shown in Fig. 5–5. The output pulse spectrum has the same shape but is now centered about zero frequency instead of f_c. The spectral components with "negative" frequencies are best interpreted as vectors rotating about the origin in the opposite direction to that of the positive frequency vectors.

The video amplifier following the detector is used to increase the pulse to an amplitude sufficient to operate the output display. Note that if the *IF* amplifier has a band width of Δf, the spectrum into the video amplifier will be $\Delta f/2$, since negative and positive frequency components are essentially the same as positive frequency components, as for example, $\cos(\omega t) = \cos(-\omega t)$.

Because of differences in target range and size, the strong radar echoes will be thousands of times greater in voltage than the weak echoes. The video amplifier cannot accommodate so great a dynamic range of signals. It is usually important that sufficient gain be used to detect the weakest targets, so the strongest targets saturate or overload the video amplifier. The last stages of amplification saturate first, with preceding stages following one by one. Very strong signals can saturate the detector and last stages in the *IF* amplifier. It is important that all *IF* and video amplifiers be designed so that saturated stages do not "block,"* but recover rapidly after a strong echo is gone, so that very weak echoes following close behind can be amplified.

The echo pulse from the video amplifier is presented on the indicator. The most popular type of presentation is Plan-Position Indication, or *PPI*, which resembles a map of the surrounding area with the radar located in the center of the indicator. At the instant the radar pulse is transmitted, the display dot starts moving radially outward from the center of the indicator. The direction of the radii are synchronized with the azimuth pointing direction of the antenna beam. Return echoes cause the dot to brighten and "paint" the target on the screen. The persistence of the indicator-tube phosphor is used to store the targets for the period between antenna scans. Many types of indicators are in use, their nature depending upon the purpose of the radar. Some of the commonest are shown in Fig. 5–9.

It has been pointed out that many echoes are usually received during each sweep of the antenna past a target. The indicator phosphor tends to integrate these multiple echoes, producing a more intense spot. The background noise is also integrated since after the amplitude-detector stage the average value of the noise is no longer zero, but it is not as effectively integrated because of

* Grid current, flowing when large signals drive grids positive, can charge capacitors sufficiently to bias tubes beyond cut-off, leaving them inoperative until the charge leaks away.

PULSE DETECTION AND INTEGRATION 91

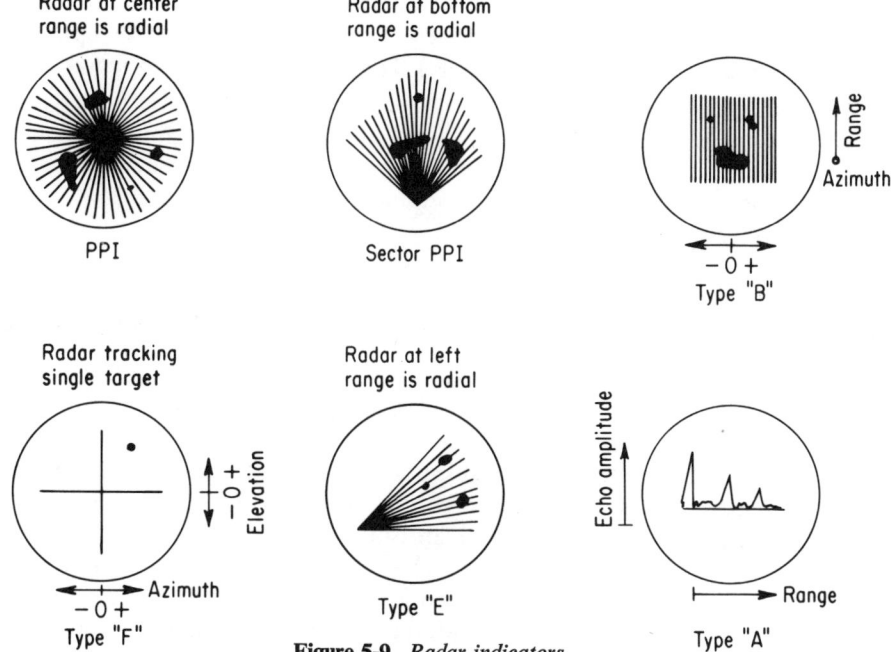

Figure 5-9. *Radar indicators.*

its random characteristics, as shown in Chapter 3. If pulse integration took place before the amplitude detector stage, the relative phases of the carrier frequency of each pulse would have to be considered when adding them. After demodulation, the carrier is removed and the pulses, which no longer have relative carrier phases in their waveforms, have only amplitudes. Post-detection integration is referred to as "non-coherent" because proper integration does not require phase coherence; it does, however, require coincidence of the envelope of each pulse atop the preceding pulses. The improvement in signal-to-noise ratio due to non-coherent integration is shown in Fig. 5–10.

Coherent pulse integration requires that the signal pulses be added so that their carriers are in phase. If N pulses of equal amplitude are so added, their sum will

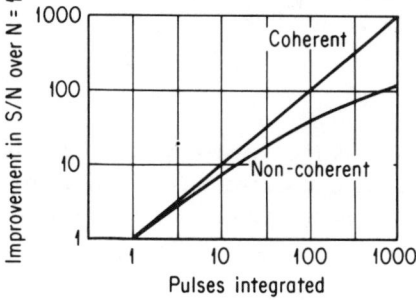

Figure 5-10. *Pulse integration.*

be a voltage N times the voltage amplitude of one pulse. The samples of noise with each pulse are completely uncorrelated since the noise is random. Therefore, it is not correct to sum the noise voltages, but it is possible to sum the noise powers. (The sum of N noise samples of equal amplitude would dissipate N times as much heat in a resistance as would one sample. However, N identical voltages added in phase would dissipate N^2 times as much power as one voltage.) Thus the noise voltage increases as $\sqrt{N}$, the signal voltage increases as N, and the signal-to-noise power ratio increases as N. As can be seen in Fig. 5–10, coherent integration offers an improvement over non-coherent integration. Methods of achieving coherent integration are discussed in the section on pulse-Doppler radars.

PULSE-RADAR RANGE

If a radar transmits a pulse at some target at a range R, the peak power in the received pulse is:

$$P_r = \frac{P_{pk}}{4\pi R^2} G_t \frac{\sigma}{4\pi R^2} A_e. \qquad (5\text{–}9)$$

P_{pk} is the peak power of the transmitted pulse, and $P_{pk}/4\pi R^2$ is the peak power density at the target if the power were radiated isotropically. G_t is the transmitting-antenna gain over an isotropic radiator, so the product of the first two terms represents the power density at the target. The radar backscatter cross section, σ, is defined as an area normal to the direction of wave propagation that "captures" all the power incident upon it and reradiates this power isotropically. It must be emphasized that this "area" is purely arbitrary and is seldom equal to any physical area of the target.

The power density at the receiving antenna is given by the product of the first three terms, and the power collected by the receiving antenna is Ae times the incident power density. Receiver noise is given by Eq. 5–4. Assuming that receiving and transmitting antennas are the same, substituting Eq. 6–5 (Chapter 6) and introducing a factor L to take care of any system losses or to account for any deficiencies in the theory, the signal-to-noise ratio out of the IF amplifier is:

$$\left[\frac{S}{N}\right]_{IF} = \frac{P_r}{P_n} = \frac{P_{pk} G_t^2 \lambda^2 \sigma L}{(4\pi)^3 kT \, \overline{NF} \Delta f R^4}. \qquad (5\text{–}10)$$

This is the conventional pulse-radar range equation. It can be modified to a more basic form by the following formulas.

The peak and average powers are related by:

$$P_{av} = P_{pk} \tau_o f_r \qquad (5\text{–}11)$$

where τ_o and f_r are as shown in Fig. 5–5. The optimum receiver band width is:

$$\tau_o \Delta f = 1.2. \qquad (5\text{–}12)$$

The antenna gain can be expressed as:

$$G = \frac{4\pi}{\psi_b} \qquad (5\text{–}13)$$

where ψ_b is the solid angle (in steradians) of the antenna beam. The effective signal-to-noise ratio after coherent integration of N pulses is:

$$\left[\frac{S}{N}\right]_{\text{eff}} = N\left[\frac{S}{N}\right]_{IF}. \tag{5-14}$$

The number of pulses received is:

$$N = T_i f_r \tag{5-15}$$

where T_i is the integration time, the time required for the antenna beam to traverse the target, and f_r is the pulse repetition rate. If the antenna beam is required to search a solid angle ψ_s in a time T_s, and spends an equal time looking in all directions within that solid angle:

$$\frac{T_i}{T_s} = \frac{\psi_b}{\psi_s}. \tag{5-16}$$

Equations 6–5 and 5–10 through 5–16 can be combined to give:

$$\left[\frac{S}{N}\right]_{\text{eff}} = \frac{\sigma A_e P_{av} T_s L}{(1.2)4\pi kT\,\overline{NF}R^4\psi_s}. \tag{5-17}$$

Notice that this range equation depends upon $P_{av}T_s/\psi_s$, the energy per solid angle, and A_e, the effective antenna area. It is independent of peak transmitted power, pulse length, pulse rate, and wavelength. The values of $[S/N]_{\text{eff}}$ required for a given probability of target detection are given in Chapter 3, (Fig. 3–8). Equation 5–17 is useful in that it indicates how various factors are interrelated and how variations in them will change the range.

It is difficult to determine the radar cross section of most targets, and as a result, accounting for the effects of changes in cross section upon radar performance becomes quite involved. Radar cross section varies with frequency, and when the target is many wavelengths in dimension it also varies greatly, or scintillates, with small changes in target aspect angle. The apparent center of reflection for a complex target can move rapidly from place to place on the target as the target slowly rotates, causing the target to appear to "jitter" back and forth in range and angle. This movement modulates the radar echoes, limiting the amount of coherent integration that can be performed.

Variations in the performance of radar operators, such as their skill, the fraction of time they can devote to watching the display, the type and quality of the display, ambient light conditions, and other factors, are almost impossible to calculate. Changes in these factors can amount to a factor of two in radar range.

It was assumed in Eq. 5–13 that the antenna gain was constant between half-power points and zero elsewhere. This assumption is usually corrected by a beam-shape loss factor of 1 to 2.5 db. It is also assumed that the antenna is not conically scanning during search and that it does not have a monopulse pattern. If these factors are present, additional loss must be considered.

If the received pulses are incoherently integrated, the loss can be compensated for as shown in Fig. 5–10. The *IF* band width should be $\Delta f = 1.2/\tau_o$ for best performance, and variations of this value must also be considered.

Still other loss factors to be considered are atmospheric absorption and rainfall, improper radar maintenance, and gradual degradation of equipment.

In practise, the range of the radar is calculated as accurately as possible, taking into account as many corrections as necessary, and then accounting for any discrepancies between calculations and flight test by adjusting the operator or field degradation factors. It is possible to use Eq. 5–17 to estimate the effects of any changes in radar parameters.

CW RADARS

The pulse radars discussed in the preceding section are often unable to detect small targets because of the echoes from the land or sea surrounding the target. This land or sea "clutter" can give echoes that are sufficiently strong and irregular that they completely obscure the target echoes. A radar

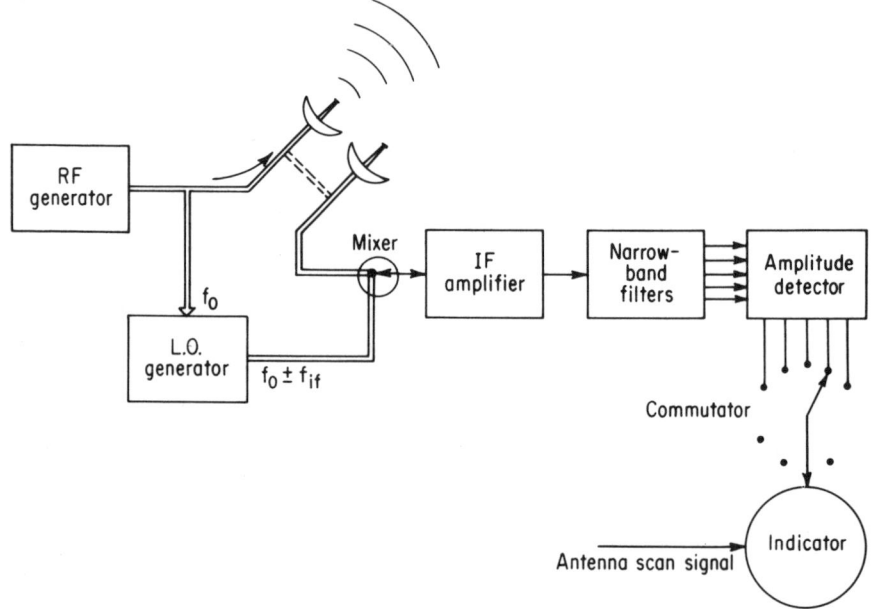

Figure 5-11. *CW radar.*

with a 2 degree antenna beam width, transmitting 1 microsecond pulses, would receive an echo from a low-flying aircraft 30 miles away but to the echo would be added the clutter from an area of ground 6000 ft wide and 500 ft long. The effective backscatter cross section of this clutter could easily be 15,000 sq ft, 150 times as large as many fighter aircraft.

One of the principal differences between many targets such as aircraft, trucks, and tanks is the velocity of the target, which results in a shift in the echo frequency due to the Doppler effect. Continuous-Wave (*CW*) radars are able to use this frequency shift to resolve moving targets. A simplified block diagram of a possible *CW* radar is shown in Fig. 5-11. The transmitter consists of a radio-frequency generator and associated power supplies. Magnetrons and klystrons are often used for the higher radar bands and conventional transmitting tubes for the lower bands. As will be seen, it is very important that the transmitter spectrum be free of any noise or spurious modulations.

Since transmission is continuous, it is almost impossible to use the transmitting antenna for receiving. A separate antenna is used, and it is oriented so as to minimize the power received directly from the transmitter. This requirement for two antennas is one of the major disadvantages of *CW* radars.

The received echoes are mixed with a local oscillator signal to produce an *IF* signal. The mixing is the same as that described for pulse radars. A local oscillator signal is generated by mixing a signal at the desired intermediate frequency with the transmitter frequency and filtering out the sum or difference component.

The *IF* amplifier requires a band width broad enough to pass the expected Doppler spread of frequencies. The mount of Doppler shift, f_d, is equal to the rate of change of round-trip radar path length expressed in wavelengths per second:

$$f_d = \frac{2v_c}{\lambda} \quad (5\text{-}18)$$

where v_c is the radial or closing velocity and λ is wavelength. At 10,000 megacycles a 2000 fps closing velocity would shift the frequency by 40 kc.

A narrow-band filter selects the Doppler-frequency component corresponding to the velocity to be detected. A bank of filters is used to detect more than one frequency simultaneously. Each of the filters is tuned to progressively increasing but overlapping frequency bands.

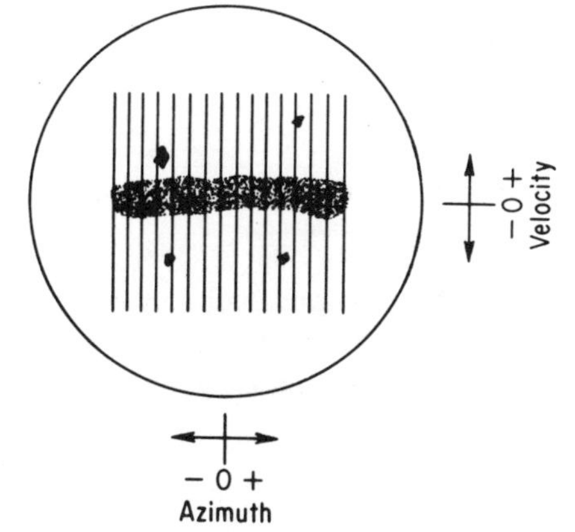

Figure 5-12. *Velocity indicator for stationary radar. The center band is due to ground clutter and spurious modulations.*

Amplitude detectors are used to demodulate the output from each filter and produce a continuous voltage that indicates the presence of a target at the velocity corresponding to the filter. A commutator is used for rapid examination of each filter detector in turn.

The principal quantities measured by *CW* radars are angles and velocity. Since velocity is not a dimension of space, no map types of display, such as

PPI, are possible. Therefore, the type of display is somewhat of a problem. Figure 5–12 shows one type of indicator that can be used. The indicator-tube spot is swept in the velocity dimension in synchronization with the commutator and is moved back and forth in the azimuth direction corresponding to the antenna motion. The commutated output from the detectors causes the spot to brighten and display the targets. The band in the center of the indicator at zero velocity is due to ground clutter and signal leakage from the transmitter.

DOPPLER-FILTER BAND WIDTH LIMITATIONS

The sensitivity of the receiver is limited by the amount of noise that passes through the narrow-band filters, the noise being proportional to the band width. For several reasons it is not possible to make the band width arbitrarily small. First, the antenna beams are scanning past the target, causing the echo-signal amplitude to vary as the two-way gain of the antennas. In other words, the continuous received signal is changed into a pulse of duration equal to the time the antenna beams spend traversing the target. The spectrum of this pulse is best explained by considering an antenna made up of an array of dipole elements instead of a reflector and feed (Fig. 5–13). As the antenna rotates, the dipoles at one end move toward the incident radar wave, while those at the other end move away. This motion results in a Doppler frequency shift along the surface of the antenna. If the antenna has a diameter, d, and an angular velocity, ω_r, (radians/sec), the Doppler variation across its surface is

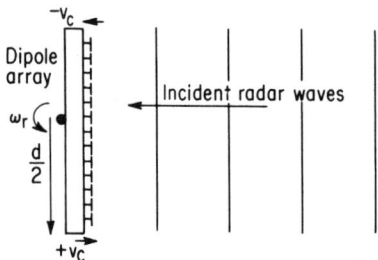

Figure 5-13. *Doppler shift due to rotating antenna.*

$$f_d = \pm 2\frac{v_c}{\lambda} = \pm \frac{\omega_r d}{\lambda}. \tag{5-19}$$

Antenna beam width from Eq. 6–6 (Chapter 6) is:

$$\theta = \frac{70}{57}\frac{\lambda}{d} \text{ radians} \tag{5-20}$$

for an antenna with tapered illumination across its aperture, and the time the beam spends on target is

$$T_i = \frac{\theta}{\omega_r}. \tag{5-21}$$

Combining Eq. 5–19 through 5–21 gives:

$$f_d = \pm \frac{1.2}{T_i}. \qquad (5\text{–}22)$$

This result, that the spectral spread is approximately the reciprocal of the pulse duration, is essentially the one to be expected when considered on a pulse basis, as illustrated in Fig. 5–5. However, as consideration of Fig. 5–13 makes clear, since the physical extent of the antenna is only $\pm d/2$, the spectrum is not spread beyond $\pm \omega_r d/\lambda$.

If a 10 kmc CW radar with a 2-degree antenna scans past a target at 60 degrees/sec, the band width should be about 36 cycles/sec for optimum S/N. To examine a 1500 fps velocity range simultaneously, 833 Doppler filters would be required. Widening the filters would reduce the number required, but it would also reduce the detection range and the velocity resolution.

A single filter could be tuned through the frequency range to be monitored, but the sluggishness of the filter must be considered. If a signal at its resonant frequency is suddenly connected across a simple RLC filter (having a half-power band width of Δf, its response will build up as $1 - \exp(-\pi \Delta t)$, reaching 63% of its final value in $1/\pi \Delta f$ sec. Assuming that a filter should not be tuned through any single frequency in less than this response time, the maximum tuning rate is

$$\frac{df}{dt} = \frac{\Delta f}{1/\pi \Delta f} = \pi \Delta f^2. \qquad (5\text{–}23)$$

This conclusion can be compared with the more accurate expression given under the discussion of receiver modes in Chapter 4, which shows that the filter response is reduced 2.3 db at a tuning rate of $\pi \Delta f^2$. It would take the 36 cps filter 7.4 seconds to examine the 1500 fps velocity range, which would require the antenna to scan at 0.27 degrees/sec. This scanning is unnecessarily slow and is wasteful of radar power. However, in applications in which maximum radar sensitivity is not required, a single swept filter is much simpler than a bank of many fixed filters.

In addition to antenna motion, target motion will also spread the spectrum of the echo signal. Yawing of an airplane, for example, rotates the wings in a way that modulates the echo in a manner similar to that of the rotating antenna previously discussed. Propeller or jet-turbine rotation can produce wide spectral spreads because of amplitude and frequency modulation.

The spectra of ground and sea returns are also spread because of the motion of leaves and waves. Some representative values of target-velocity spreads are given in Table 5–1.

System instabilities such as modulations of the transmitter or local oscillator signals because of power-supply variations or mechanical vibrations can degrade CW radar performance by producing spurious sidebands that

98 RADAR CONSIDERATIONS

Table 5-1. *Typical velocity spreads for various distributed targets*

TARGET	VELOCITY SPREAD
Sparsely wooded ground, calm air	± 0.02 fps
Heavily wooded ground, 20 mph wind	± 1.0 ,,
Sea, windy day	± 3.5 ,,
Chaff	± 4.0 ,,
Rain cloud	± 7.5 ,,

fall inside the expected Doppler band width. This system noise enters the receiver by direct leakage from the transmitter, or as modulation on the clutter return. Amplitude modulation at a single frequency f_m generates two sidebands displaced by f_m above and below the carrier frequency. If the percentage modulation is α, each sideband power is $25 \times 10^{-6} \alpha^2$ times the carrier power. Thus, 1% modulation sidebands are 46 db below the carrier.

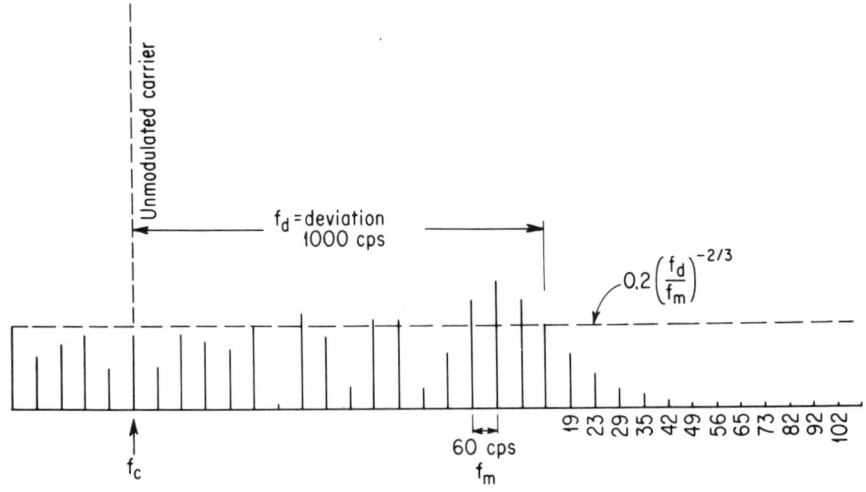

Figure 5-14. *FM spectrum for a 1000 cps deviation at a 60 cps modulation frequency.*

Frequency modulation, wherein the carrier frequency deviates $\pm f_d$ at a rate of f_m, results in approximately $2 f_d/f_m$ major sidebands spaced every f_m as shown in Fig. 5-14. Individual sideband amplitudes vary considerably over the range $\pm f_d$, but the majority of the amplitudes are within a factor of four of:

$$P_{SB} \approx 0.2 \, P_c \left(\frac{f_d}{f_m}\right)^{-2/3} \qquad (5\text{-}24)$$

where P_{SB} and P_c are sideband and unmodulated carrier powers, respectively. (Sidebands seldom exceed four times this result.) For example, if the carrier deviates ±1 kc at 60 cps, the sidebands are about 15 db below the unmodulated carrier power and extend 1 kc either side of the carrier frequency. It must be emphasized that the "carrier," as such, has "disappeared," and is now no stronger than the sidebands. The true carrier is reduced to half-power when f_d/f_m is 1.1, and is reduced by 10 db when f_d/f_m is 2.2. Thus, if the transmitter deviation is not kept small compared with the Doppler-filter band widths, only a small portion of the available power will pass through the narrow-band filter, an obvious waste of radar power. Also, as can be seen from the many spectral lines in Fig. 5-14, too narrow a filter would have many ambiguous possible Doppler velocities to measure, limiting velocity accuracy.

When a filter is wider than either the frequency deviation or the modulation frequency, so that it effectively passes all the major spectral lines, the

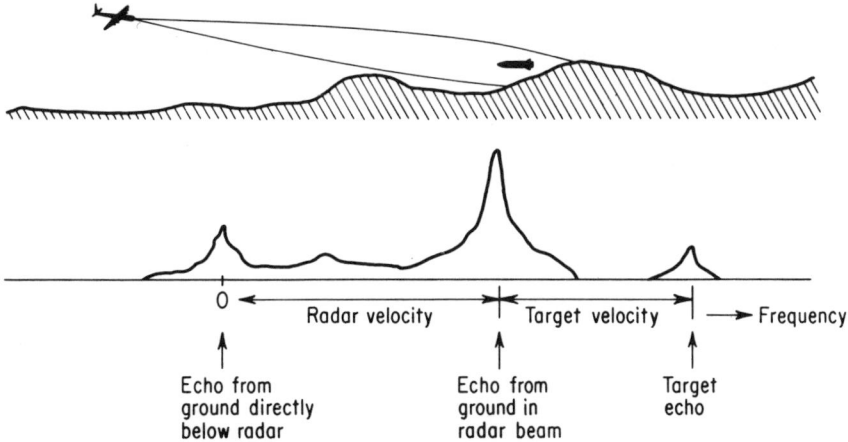

Figure 5-15. *Doppler spectrum from ground and target.*

filter passes the signal with little degradation because of the FM. A filter this wide may be too wide to meet the band width requirements for optimum coherent integration. Excessive noise is then passed by the filter. This noise can be reduced by inserting a low-pass network after the amplitude detector following the filter. A simple *RC* ladder network is often used. The network has the effect of providing the additional integration lacked by the Doppler filter. This is non-coherent integration and is not quite as effective as coherent integration, as explained in Chapter 3. Of course the basic velocity resolution and accuracy are determined by the band width of the Doppler filter, not the *RC* network.

An airborne *CW* radar receives a wide range of Doppler frequencies, as shown in Fig. 5-15. Even with the antenna pointed forward, some return from the ground enters the receiver through the antenna sidelobes, or possibly by scattering from aircraft or radome parts near the antenna. Of course there are very large returns from the ground illuminated directly by the antenna beam. The Doppler on this ground clutter is proportional to the closing rate between the radar and each small reflecting element of ground. The closing velocity differs for various elements over the area illuminated, resulting in a spreading of the clutter frequency. Except for clutter fluctuations of the sort listed in Table 5-1 (and possible antenna motion), there would be no clutter Doppler frequencies greater than those corresponding to the forward velocity of the aircraft, if there were no modulation of the transmitter.

The FM spectrum given in Fig. 5-14 shows the spectral lines disappearing at frequencies slightly outside the range of frequency deviation ($\pm f_d$). This apparent reduction to zero can be misleading since in theory the lines extend over an infinite range on either side of the carrier. It is important to have some idea of how rapidly the modulation spectral components die away so they can be compared with the amplitudes of possible target returns they might obscure. The following equation gives a good estimate of the power of any sidebands outside the deviation range:

$$P_{SB} = P_c \frac{e^{2a}}{2\pi a} \left(\frac{b}{n+a} \right)^{2n} \quad (5\text{-}25)$$

where:
$$a = \sqrt{n^2 - b^2}$$

$$n = \frac{f_n - f_c}{f_m}$$

$$b = \frac{f_d}{f_m}$$

where $f_c + f_d < f_n < f_c - f_d$, and f_c, f_d, f_m and f_n are the carrier, deviation, modulation, and sideband frequencies respectively, P_{SB} is the sideband power and P_c is the unmodulated carrier power. As an example, the amplitudes of some of the sidebands are indicated in Fig. 5-14 by numbers giving their reduction from the unmodulated carrier in decibels.

Since the local oscillator signal is generated by mixing a signal at the *IF* with the transmitter, any variations in the transmitted frequency will also be in the local oscillator signal. Since the receiver mixer subtracts the local oscillator and echo-signal frequencies, the *IF* signal would not contain any of the frequency modulation if the echo were not delayed by the propagation time.

Figure 5-16 shows two sinusoidal frequency variations, one delayed from the other by τ. The difference between these two variations is also shown. Since both variations must be at the same frequency, ω_m, their difference also has a frequency of ω_m. It can also be deduced by inspection that the maximum amplitude the difference can have is twice the original variation, and that the difference is delayed by $(3\pi/2) + \tau/2$ radians. Thus, if the original variation in transmitter frequency is given by

$$\omega_d \sin \omega_m t \qquad (5\text{--}26)$$

the IF variation is:

$$2\omega_d \sin\left[\frac{\omega_m \tau}{2}\right] \cos\left[\omega_m\left(t - \frac{\tau}{2}\right)\right]. \qquad (5\text{--}27)$$

For example, at a target range of 25 miles a transmitter frequency deviation of ± 18 kc at a rate of 60 cps is reduced to an IF variation of only ± 1 kc

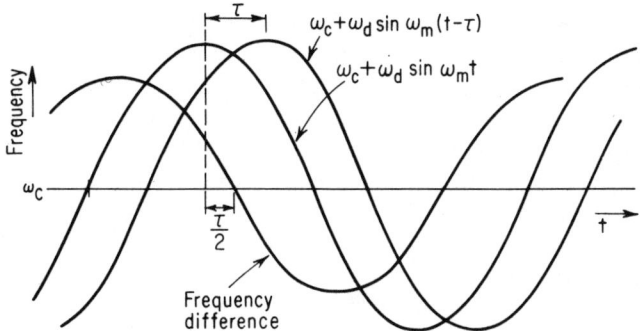

Figure 5-16. *FM in a CW radar.*

by this effect. However, if the modulation occurred at 1600 cps instead of 60 cps, the deviation would be doubled.

Frequency modulation is sometimes used in *CW* radars to obtain target range. The echo signal is amplitude-limited so it will be independent of target size, then it is compared in phase with the transmission modulation, generating a voltage proportional to $\sin(\omega_m \tau/2)$. This determines target range, since τ is proportional to range.

To illustrate the effects of clutter on *CW* radar performance, consider the following example: an aircraft at 24,000 ft altitude is flying 1000 fps, searching for a target that is 25 miles ahead and flying close to the ground. Its *CW* radar has the following characteristics:

Power	200 w
Frequency	10,000 mc
Antennas	2 ft diameter
	3.5° beam width
	33 db gain
Noise figure	10 db
Doppler filters	2 kc band width each.

The ground area illuminated ahead by the antenna beams is an ellipse 9.2 miles long and 1.5 miles wide, covering an area of 400 million sq ft. The total

Doppler spread over this area is only 200 cps, which is small compared to the 2 kc filter band width. The effective back-scattering area for ground clutter is roughly 1 % of the area illuminated. Taking the effective back-scattering area as 4 million sq ft, the received main-lobe clutter power is 3×10^{-11} w. The receiver noise in the 2 kc band width is 8×10^{-17} w, which is 56 db below the clutter level. If the spurious frequency modulation of the transmitter is ± 18 kc at 60 cps, the clutter spectrum will be approximately the same as illustrated in Fig. 5–14. The target spectrum is also similar to the one shown in Fig. 5–14 except that it is 41 db below the clutter level (for a target cross section of 325 sq ft), and displaced from the clutter Doppler by the target Doppler. If the target Doppler is separated from the clutter Doppler by 3 kc or more, all the significant target spectral components will be stronger than the clutter components at the same frequencies and the clutter will not interfere with the target echo. For a target-clutter Doppler separation of 1.5 kc, the lower half of the target spectral components would be obscured by the upper clutter components. Assuming a target velocity of 500 fps and a minimum target-clutter Doppler separation of 2 kc, then there is a 23 degree sector centered about a line normal to the target's direction of flight within which the radar cannot see the target because of ground clutter. Likewise, a ground vehicle slower than 70 mph in the direction of the radar would be obscured by clutter.

If the spurious modulation of transmitter is reduced so that its spectral lines are confined to a narrower band width, the clutter becomes less of a problem. Targets with positive-Doppler frequencies can be slower and still be sorted from the clutter. Narrower Doppler filters can be used. If the filters are narrower than the Doppler spread due to variations of ground velocity over the illuminated ground area, they will pass only a portion of the total power received within the beam width. This is analogous to the pulse-radar situation, where a narrow pulse reduces ground clutter because only a portion of ground area covered by the antenna beam is resolved at any instant. Then the velocity resolution of a moving airborne *CW* radar can be used to effectively increase the radar ground mapping resolution, beyond that provided by the antenna beam, just as a narrow pulse increases radar resolution in range.

CW RADAR RANGE

The basic detection range of a *CW* radar is quite easily calculated, using the same reasoning that was developed in Eq. 5–9, except that average power is used instead of peak power:

$$P_r = \frac{P_{av} G_t \sigma A_e}{(4\pi)^2 R^4}. \qquad (5\text{–}28)$$

The receiver noise is given by Eq. 5–4, and the signal-to-noise ratio is then

$$\frac{S}{N} = \frac{P_{av} G_t \sigma A_e L}{(4\pi)^2 kT \overline{NF} \Delta f R^4}. \qquad (5\text{–}29)$$

The factor L is introduced to account for any system losses or deficiencies in the theory.

This is the signal-to-noise ratio after the narrow-band Doppler filter, which has a band width of Δf. The optimum value of Δf is $1.2/T_i$, where T_i is the time the antenna beam spends scanning past the target.

Using Eqs. 5-4, 5-13 and 5-16, which apply to all radars, gives:

$$\left[\frac{S}{N}\right]_{\text{eff}} = \frac{\sigma A_e P_{av} T_s L}{(1.2)\, 4\pi\, kT\overline{NF}R^4 \psi_s}. \tag{5-30}$$

This equation is the same as the range equation (5-17) developed for pulse radars.

PULSE-DOPPLER RADARS

Pulse radars give nice, maplike presentations of the surrounding terrain and targets, but the targets are often obscured by the ground return. Trucks, tanks, and fighter aircraft seldom have radar cross-sections large enough to compete with the echo from the large ground area within the antenna beam.

Pure CW radars can take advantage of the Doppler from moving targets to discriminate them from the clutter, but are unable to provide range resolution. They provide velocity resolution and can use FM techniques to measure

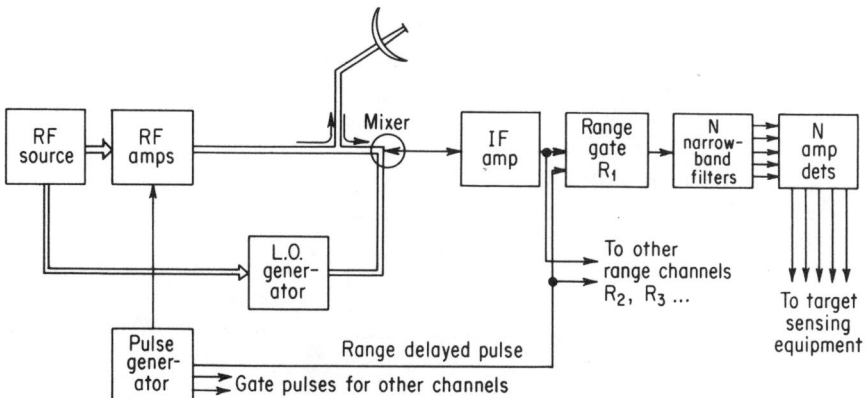

Figure 5-17. *Pulse-Doppler radar.*

range on a single target, or multiple targets if each is resolved by velocity. However, an approaching group of aircraft spread out in range and all having the same velocity could not be distinguished from one another.

It is possible to combine the functions of pulse and CW radars so as to simultaneously resolve in both range and velocity. A block diagram for such a radar is shown in Fig. 5-17. The transmitter generates narrow pulses for transmission, like a pulse radar. It must also develop a coherent local oscillator signal, and in this respect is like a CW radar. This coherence is

104 RADAR CONSIDERATIONS

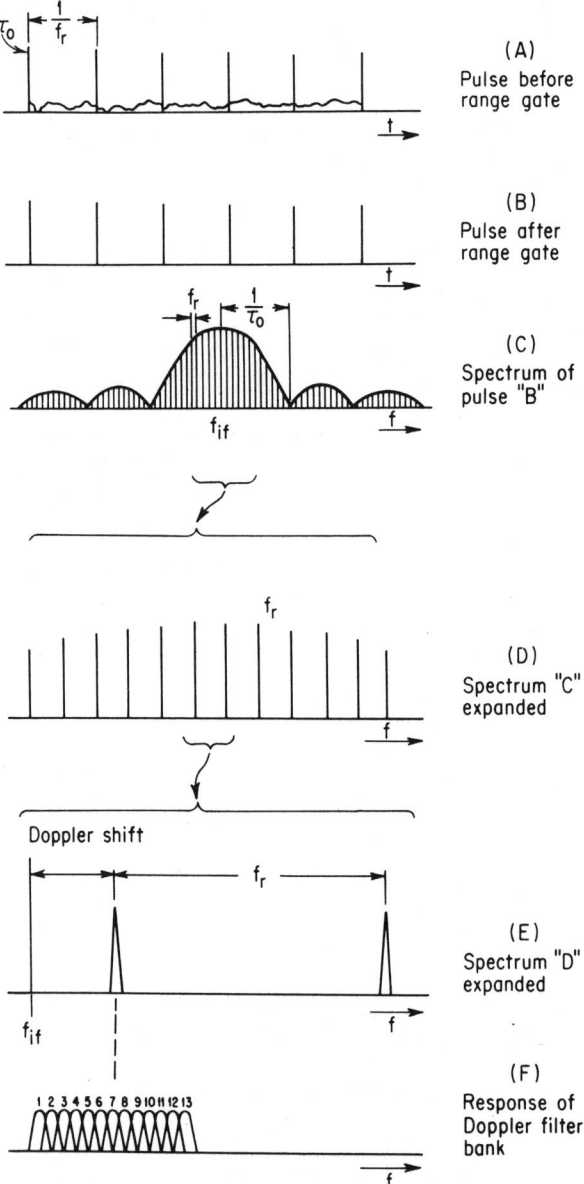

Figure 5-18. *Pulse-Doppler filtering.*

often obtained by amplifying the signal from a stable *CW* source and then pulsing the signal supplied to the *RF* power amplifiers.

The received echo pulse is mixed with the local oscillator signal to produce the *IF* signal. The *IF* echo pulse is passed through a range gate that is triggered "open" during the time the pulse is expected. It is "closed" between pulses and thus shuts out the receiver noise during this period.

Figure 5–18 shows the pulse waveforms and spectra. Figure 5–18(B) shows the removal of the noise between pulses present in Fig. 5–18 (C). The spectrum of the waveform in (B) is shown in (C) and is the same pulse spectrum illustrated in Fig. 5–5. Figures (D) and (E) are expansions of the frequency scale, which illustrate the individual spectral components of the pulse. Since the pulse envelope is purely periodic and is imposed upon a *CW* carrier (no spurious carrier AM or FM), all the components of its spectrum are pure *CW*. If the target is moving toward the radar, the frequencies of all spectral components are shifted by the Doppler, as shown in (E). One of the central spectral components can be regarded as the signal in a *CW* radar and can be processed as such. Figure (F) shows an array of narrow-band Doppler filters used to sort out the target component, which happens to pass through "filter number 6." It is obvious from Figs. (D), (E), and (F) that the pulse-repetition period must exceed the Doppler range of possible targets so that no more than one spectral line can pass through the filters and produce velocity ambiguities.

It is not readily apparent that filtering out only one spectral line and throwing away all the rest will give satisfactory radar sensitivity. If E_c is the peak voltage of the pulse, the voltage amplitude of the central line is $\tau_o f_r E_c$. The power in this line is then only $(\tau_o f_r)$ times the *average* pulse power. What has happened is that the range gating has also reduced the noise-power density by a factor $\tau_o f_r$, so that the signal-to-noise ratio into the narrow-band Doppler filter is in the ratio of average signal to average noise, as for a *CW* radar.

To compute the noise out of the range gate, the gating action can be regarded as multiplying the *IF* amplifier output by a function that is unity during the pulse interval τ_o and zero between pulses. This function is the same as the one shown in Fig. 5–5. Each of its spectral lines can be regarded as individual "local oscillators," and the multiplication is the equivalent of mixing each of these "local oscillators" with the noise and summing the results.

The amplitudes of the local oscillator signals are:

$$\tau_o f_r \frac{\sin x}{x} \tag{5-31}$$

where $\qquad x = (f - f_c)\pi\tau_o$

and the mixer output voltages (*RMS* volts per cycle band width) are:

$$\frac{1}{2} D_n \tau_o f_r \frac{\sin x}{x} \tag{5-32}$$

where D_n is the noise voltage per cycle out of the *IF* amplifier. The *IF* bandpass will be assumed constant between f_a and f_b and zero elsewhere. The noise voltages have random relative phases and thus cannot be coherently added, but their powers can be. The resultant noise-voltage density D_r is:

$$D_r^2 = \sum_{f=f_a}^{f_b} \left[\frac{1}{2}D_n\tau_o f_r \frac{\sin x}{x}\right]^2$$

$$\simeq \left(\frac{1}{2}D_n\tau_o f_r\right)^2 \frac{1}{\pi f_r \tau_o} \int_a^b \frac{\sin^2 x}{x^2} dx \qquad (5\text{--}33)$$

$$\simeq \left(\frac{1}{2}D_n\right)^2 \tau_o f_r \frac{2}{\pi} \int_0^b \frac{\sin^2 x}{x^2} dx$$

$$b = -a = \frac{\pi}{2}\tau_o(f_b - f_a).$$

When the *IF* band width is large:

$$D_r^2 = (\tfrac{1}{2}D_n)^2 \tau_o f_r. \qquad (5\text{--}34)$$

The factor $\tfrac{1}{2}$ results from the mixing, as indicated in Eq. 5–3.

The effect of the range gate upon the signal pulse can also be computed in a similar manner. If E_c is the peak pulse voltage, the voltage amplitude of the central spectral component is

$$E_s = \sum_{f=f_a}^{f_b} \frac{1}{2}\left(\tau_o f_r \frac{\sin x}{x}\right)\left(\tau_o f_r E_c \frac{\sin x}{x}\right)$$

$$= \frac{1}{2}E_c(\tau_o f_r)^2 \frac{1}{\pi \tau_o f_r} \int_a^b \frac{\sin^2 x}{x^2} dx \qquad (5\text{--}35)$$

$$= \frac{1}{2}E_c \tau_o f_r \frac{2}{\pi} \int_0^b \frac{\sin^2 x}{x^2} dx.$$

The signal-to-noise density power ratio after gating is:

$$\frac{S}{N_d} = \left(\frac{E_s}{D_r}\right)^2 = \left(\frac{E_c}{D_n}\right)^2 \tau_o f_r \frac{2}{\pi} \int_0^b \frac{\sin^2 x}{x^2} dx. \qquad (5\text{--}36)$$

The function

$$\frac{2}{\pi}\int_0^b \frac{\sin^2 x}{x^2}$$

is plotted in Fig. 5–6, showing that it increases as the *IF* band width increases. This increase indicates that the *IF* band width can be somewhat wider than $\tau_o \Delta f = 1.2$, the optimum value derived for pulse radars. After all, it is the Doppler-filter band width that will restrict the noise. For this case

$$\frac{S}{N_d} = \left(\frac{E_c}{D_n}\right)^2 \tau_o f_r. \qquad (5\text{--}37)$$

Since $(E_c/D_n)^2$ is equal to the ratio of peak pulse power to RMS noise-power density, and $\tau_o f_r$ reduces peak power to average power, the output S/N_d after gating is equal to the ratio of the average signal power to noise density before the gate. The average signal power before the gate is (from Eq. 5–9):

$$P_r = \frac{P_{av} G_t \sigma A_e}{(4\pi)^2 R^4}$$

and the noise-power density is $kT\overline{NF}$, therefore after gating:

$$\frac{S}{N_d} = \frac{P_{av} G_t \sigma A_e}{(4\pi)^2 R^4 kT\overline{NF}}. \tag{5-38}$$

The Doppler-filter band width is Δf cycles, and the signal-to-noise power ratio following it is:

$$\frac{S}{N} = \frac{S}{N_d}\frac{1}{\Delta f} = \frac{P_{av} G_t \sigma A_e}{(4\pi)^2 kT\overline{NF}\Delta f R^4}. \tag{5-39}$$

The optimum filter band width is

$$\Delta f = \frac{1.2}{T_i}$$

where T_i is the time required for the antenna to scan past the target. Substituting Eqs. 5–16 and 5–13 gives:

$$\frac{S}{N} = \frac{P_{av} \sigma A_e T_s L}{(1.2) 4\pi kT\overline{NF} R^4 \psi_s} \tag{5-40}$$

which is the same as the radar-range equations developed for pulse and CW radars.

The range gate shown in Fig. 5–17 allows the bank of N Doppler filters to quantize one range interval into N possible velocity values. If M range intervals are to be examined simultaneously, M range gates and Doppler-filter banks are required. The total number of Doppler filters is M times N, which can be inconveniently large. As mentioned concerning CW radars, the number of filters can be reduced by making them wider than optimum and inserting low-pass networks after the amplitude detectors to narrow the effective band width.

Since the pulse-repetition frequency must be greater than the maximum Doppler, the number of range gates required is less than would be needed in a conventional pulse radar. For example, to accommodate a 4000 fps target closing velocity, a 10,000 mc radar would need a repetition frequency of more than 80 kc. Assuming 1 microsecond pulses, the maximum possible number of range gates would be about eleven. An 80 kc repetition rate results in range ambiguities every 6250 ft. Such ambiguities can be resolved by changing the repetition rate among several different values and noting the shift in position of the echo pulse in relation to the transmitted pulse, or by coding the transmissions and measuring the round-trip propagation delay.

As far as any single Doppler filter is concerned, since it sees only a single spectral line, the frequency stability and spurious modulation requirements upon the transmitter are the same as for a CW radar. The effects of antenna motion, ground or sea motion, and target modulations are also the same as described for CW radars.

Pulsing the RF amplifiers introduces additional complications as far as purity of spectrum is concerned. Any variations in the shape of the pulse envelope from pulse to pulse, or any phase modulation of the amplifiers that is not the same for each pulse will spread the spectral lines. In short, unless all the pulses are exactly the same, the transmitter output is not strictly periodic and cannot be represented by individual spectral lines spaced f_r apart.

This need for coherence can be illustrated by considering what would happen if a magnetron were used for the transmitter. A magnetron is off between pulses, and during the time the modulator pulses it on, its oscillations can be described as $\sin(\omega t + \phi)$. Although ω is the same for each pulse, the relative phase, ϕ, has a new value for each pulse and is random from pulse to pulse. Some idea of how this change affects the spectrum can be had by assuming that for a set of n pulses, ϕ can have n random values, but that each set of n pulses is the same. Then, if the original pulse waveform had a period of $1/f_r$, the new period is n/f_r. Now the resulting spectral lines are spaced in frequency by f_r/n, and are reduced to $1/\sqrt{n}$ times the original voltage. As n is made very large, the spectrum becomes essentially continuous, and reference to Fig. 5–18(E) shows that such a spectrum is useless as far as Doppler filtering is concerned. For this reason, phase coherence of the IF signal must be maintained from pulse to pulse. It is possible for magnetrons to meet this requirement by locking the phase of a stable oscillator with the phase of each magnetron pulse, and using this oscillator as a reference for developing the Doppler frequency. This method is not practical when more than one pulse is in transit at any one time.

Having to chop narrow pulses from a continuous RF source makes it more difficult for a pulse-Doppler radar to maintain a pure spectrum than for a CW radar. However, a pulse-Doppler radar, since it does transmit narrow pulses, resolves only a narrow strip in range (one or more, depending on the distance between pulses in relation to the ground area illuminated), and receives correspondingly less clutter than a CW radar that resolves only in velocity; it often receives its clutter from the entire ground area illuminated by the antenna. Any clutter rejection a CW radar can accomplish on the basis of velocity resolution can also be obtained in the same manner by pulse-Doppler.

Pulse-Doppler radars often require higher pulse-repetition frequencies than pulse radars, and thus for a given average power they have lower peak pulse powers. This characteristic reduces the problem of waveguide and antenna power breakdown that often limits pulse radars.

Since pulse-Doppler radars do not transmit and receive simultaneously, they do not have the transmitter-receiver isolation problems of *CW* radars. However, their higher pulse rate means that they cannot wait until echoes from all ranges of interest have returned, as pulse radars can, which results in "blind" ranges (ranges for which the echo arrives during transmission). This difficulty may not be significant if the pulses are short and the target closing rate is high. It can be avoided by switching the pulse-repetition rate.

MOVING TARGET INDICATION

Conventional pulse radars can be modified to a form of pulse Doppler that provides enhancement of moving targets, or Moving Target Indication (*MTI*). A block diagram for such a pulse *MTI* radar is shown in Fig. 5–19.

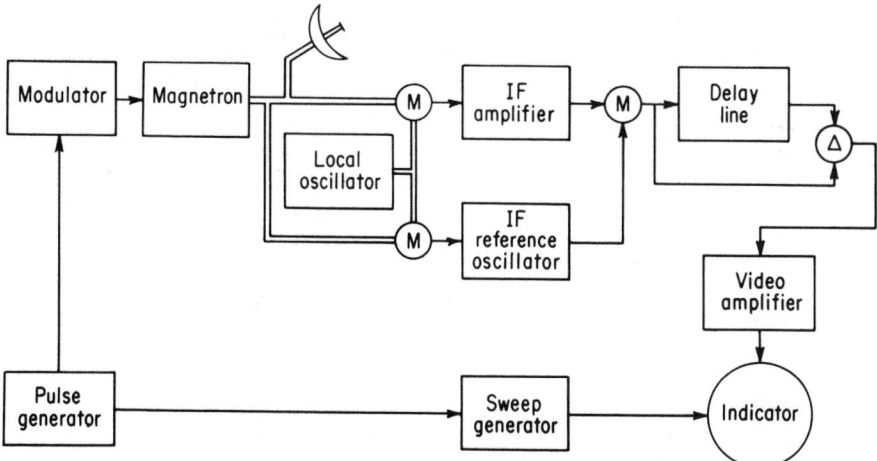

Figure 5-19. *Radar with moving target indication.*

In addition to the usual *IF* amplifier channel, the local oscillator is mixed with the magnetron pulses and the resulting *IF* pulses are used to synchronize the phase of an *IF* reference oscillator. When the reference is mixed with the *IF* amplifier output, the resulting difference frequency is modulated by the Doppler. The *IF* pulses can be represented by

$$\left\{ \text{⊓⊓⊓} \right\} \times \cos(\omega_{if} + \omega_d)t \qquad (5\text{–}41)$$

where ω_{if} and ω_d are the *IF* and Doppler frequencies. Multiplying this composite waveform by the reference signal, $\cos(\omega_{if})t$, and taking the difference frequency gives:

$$\left\{ \text{⊓⊓⊓} \right\} \times \tfrac{1}{2}\cos(\omega_d t) \qquad (5\text{–}42)$$

Thus, the *IF* pulses are changed into video pulses that vary in amplitude as cos $(\omega_d t)$. Since adjacent pulses differ in amplitude, if each pulse is subtracted from the pulse preceding it by means of a delay line followed by a comparator, a difference or "residue" is obtained. If the target is stationary there will be no Doppler, all pulses will be of the same amplitude, and the residue will be zero. Thus pulses from moving targets are enhanced while stationary clutter is rejected.

Another interpretation of this *MTI* enhancement can be had by examining the frequency response of the delay line and comparator, which is obtained by connecting a sine-wave oscillator to the input and observing the comparator output as a function of the oscillator frequency. The result is shown in Fig. 5–20. Any video pulses from stationary targets will have spectral lines that occur at 0, f_r, $2f_r$, and so on, and thus will not pass through the *MTI* filter. Moving targets will have spectral lines shifted in frequency by the Doppler, f_d, $f_r + f_d$, $2f_r + f_d$, and so on. These will be passed by the filter as long as f_d is not equal to a multiple of f_r. The delay line comparator is a convenient scheme for obtaining a rejection filter for each line of the clutter spectrum. As can be seen from the response shown in Fig. 5–20, the rejection notch is narrow and the clutter rejection is reduced if the clutter spectral lines are spread by antenna rotation, vegetation or sea motion, or pulse-to-pulse fluctuations in magnetron output.

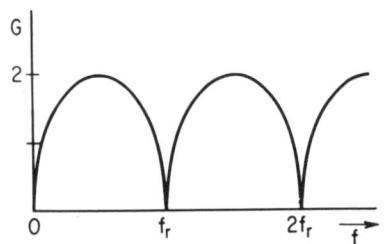

Figure 5-20. *Frequency response of MTI unit.*

If the radar is airborne, some arrangement must be made to compensate for the Doppler due to aircraft velocity. This problem can be met by shifting the *IF* reference oscillator frequency by the amount of the Doppler. Another way is to consider the stationary clutter echoes as the reference signal and regard the *IF* signal as the sum of the reference (the clutter) and moving target echo. An amplitude detector after the *IF* amplifier mixes these two signals and produces the Doppler frequencies. This clutter-referenced or non-coherent *MTI* is simpler because it does away with the reference oscillator and mixer.

The *MTI* filter has some effect upon the radar detection range. If the Doppler frequency, f_d, shifts the spectral lines into the region of peak gain ($G = 2$), the echo pulses will be doubled in voltage, and the pulse power will be increased four times. The two receiver noise signals into the subtraction unit are uncorrelated because the delay time is much greater than the fluctuation rate of the noise (approximately the reciprocal of the noise band width). Therefore, the noise power out of the subtractor is the sum of powers of the two noise inputs. The result is an overall maximum increase in *S/N* of

3 db due to the coherent integration of two signal pulses. Of course, if the Doppler is such that the spectral lines do not occur at the maximum gain region of the *MTI* filter, this increase will be less.

It is apparent from Fig. 5–20 that whenever the Doppler frequency is equal to the pulse-repetition frequency, or some multiple of it, the radar cannot see the moving target. For example, a radar operating at 10,000 mc, with a *PRF* of 1000 pulses/sec, would reject the echo of a truck moving toward it at a speed of 34 mph (or 68 mph). The pulse-repetition frequency is not easily changed, since this alteration would require a new delay line system.

TARGET TRACKING SYSTEMS

The first operation a radar performs is to search the volume of space in which the target is expected and locate its position in radar coordinates, which are usually some combination of elevation and azimuth angles, range, and velocity. When the target has been located the radar is used for tracking purposes. Sometimes the radar scans continuously, reporting its observations to a computer that keeps track of targets and anticipates their positions on the next scan. This "track-while-scan" technique provides multiple-target tracking with a single radar. However, fire-control radars are usually concerned with a single target at any one time, and the control problem often needs the higher target data rate provided by continuous tracking. The radar search operation is primarily intended to determine the presence of the hostile vehicle. The emission of false and confusing signals during this search mode does nothing to conceal the presence of the vehicle from the radar. On occasion, however, noise jamming has been misinterpreted by inexperienced radar operators as a degradation of the equipment rather than hostile radiations. For the most part, interfering with the normal tracking operation of a radar has proven to be an effective counter-measure technique. Any consideration of such ECM requires a knowledge of radar tracking systems.

ANGLE TRACKING

When an antenna is pointed directly at a target, the antenna gain is maximum, but the rate of change of gain with angle is zero, which makes it difficult to determine the exact angular position of the target within the beam. One way around this difficulty is to rotate the beam in a circle so that the beam centerline generates a cone, as shown in Fig. 5–21. Any target on the axis of the cone will be equidistant from the beam axis at all times, and the antenna gain will be constant (providing, of course, that the beam pattern is symmetrical about the beam axis). If the target does not lie on the cone axis, the antenna gain will vary at the rotational rate, modulating the amplitude of the return signal. The amplitude of the modulation depends upon the amount the target is off the cone axis and the phase of the modulation determines the direction. A reference signal is generated by the mechanical scanner and compared in phase with the signal from the amplitude detector.

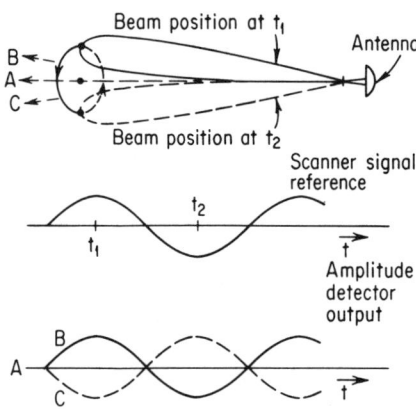

Figure 5-21 shows three different target directions, on axis (A), above axis (B), and below (C) axis, and the resulting amplitude detector outputs.

A block diagram of a typical conical-scan angle tracking system is shown in Fig. 5-22. Two phase detectors are used, one detecting the in-phase, or elevation component of modulation, and the other detecting the quadrature, or azimuth component. The tracking-error signals from the phase detectors are used to position the antenna tracking servos so that the cone axis tracks the target.

Figure 5-21. *Angle tracking by conical scanning.*

So that the phase detector outputs will not be a function of target size or range, automatic gain control (*AGC*) is used to adjust *IF* amplifier gain

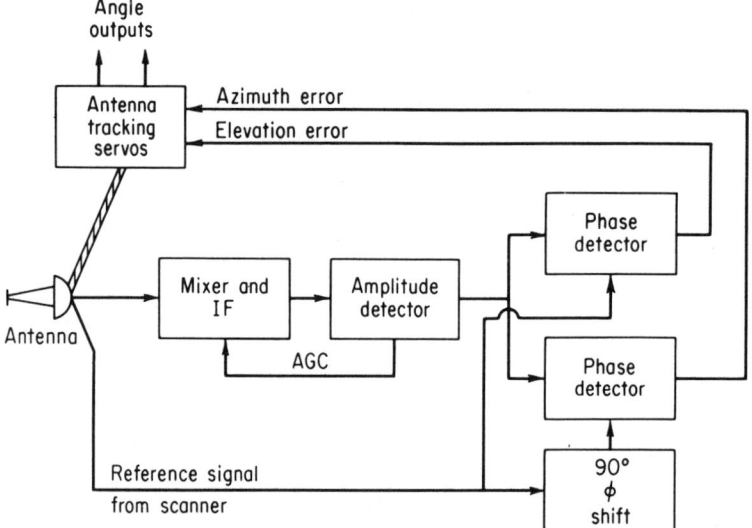

Figure 5-22. *Block diagram of conical-scan tracking system.*

and keep the *average* echo-signal amplitude from the amplifier constant. The *AGC* response must not be so fast that the conical-scan modulation is removed.

Any amplitude modulations of the echo at the conical-scan frequency that are produced by the target will introduce tracking errors into the servo system. Such modulations may be unintentionally caused by such things as propellers, or they may result from carefully designed ECM repeaters that sense the radar scan frequency and generate a strong echo signal having the same modulation frequency, but with the modulation so phased as to cause the tracking servos to move the antenna away from instead of toward the target.

Some of the amplitude-variation problems that affect conical-scan systems can be avoided by using Simultaneous Lobe Comparison (*SLC*). Such a system is shown in Fig. 5–23. A composite antenna forms upper and lower beams by placing offset feeds in a single reflector. Since the antenna patterns

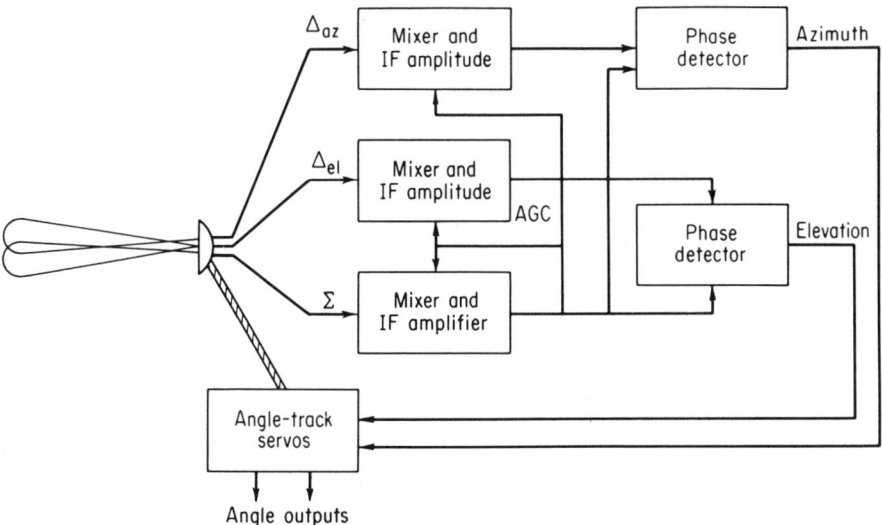

Figure 5-23. *Angle tracking by simultaneous lobe comparison.*

are identical, halfway between the two-beam axis the antenna gains and phases are equal. Any variation on either side of this mid-beam axis will cause the amplitudes or phases of a target to be unequal between the two channels. The feed waveguides are arranged so that sum (Σ) and difference (Δ) signals are formed from these beams. The difference signal is compared with the sum reference by the phase detector to generate an angle error voltage for the antenna servo. When a target is off the mid-beam axis, the difference signal is not zero and the servos drive the antenna to position the mid-beam axis on the target, reducing the error to zero. Four beams are usually used, and differences are developed between the upper and lower sets and the left and right sets to position the antenna in both elevation and azimuth. The sum consists of all four beams.

Amplitude variations in the echo will be identical in all four beams. The sum reference signal is used to control the gain of the sum *IF* amplifier so that the sum *IF* output voltage is constant. The same control voltage is used to adjust the gain of the two difference amplifiers so that all three *IF* amplifiers have the same gain. Then

$$GS_i = S_o = C \text{ (constant)} \qquad (5\text{--}43)$$

$$GD_i = D_o, \qquad (5\text{--}44)$$

where G is amplifier gain, S_i, S_o, D_i, D_o are the input and output voltages of the sum and difference amplifiers. Since the gains are equal

$$D_o = C\frac{D_i}{S_i}. \qquad (5\text{--}45)$$

Thus the *AGC* has the effect of forming the ratio of the sum and difference signals, and since any amplitude variations are identical in D_i and S_i, they do not appear in D_o.

Simultaneous lobe comparison tracking is immune to amplitude variations of the target echoes and works well with either pulse or *CW* radars. This system has the ability to track signals that are not radar echoes, such as pulse or noise signals from enemy jammers, and can develop angle information from a single pulse. Basically, it locates the target direction by sensing the angle of arrival of the radiation and aligns the antenna axis normal to the plane of the phase fronts.

RANGE TRACKING

Range tracking gates turn on the radar receiver at the instant a target return is expected and leave it off during the remaining time. Since the target is usually moving, the range gate has to be continually repositioned in range or the target will move out of the gate and be lost. Automatic range tracking is usually accomplished by a "split-gate" system, such as the one shown in Fig. 5–24. Two gates are generated, the "early" and the "late" gates, and they are positioned in time so that half the echo pulse passes through each gate. If the pulse is not centered between the gates, one gate output is larger than the other, and the difference between the amplitude detectors is used as an error voltage to adjust the time at which the gate generator produces the gate pulses. When the split gates are properly adjusted the error voltage will be zero. Often, the gate generator uses the error voltage to control the range rate, or velocity, of the gates in addition to positional control. This control provides the gates with velocity "memory," so that if the target echo should fade, they will continue to move at the last target velocity until the echo reappears, reducing the chances of the target slipping outside the gate.

Amplitude variations are removed from the echo pulse used for range tracking, either by *AGC* or by limiting, so that the error signal is independent of target size or range. The gate generator also provides a gating pulse that is centered about the echo pulse and is used to operate any other range gates

RANGE TRACKING 115

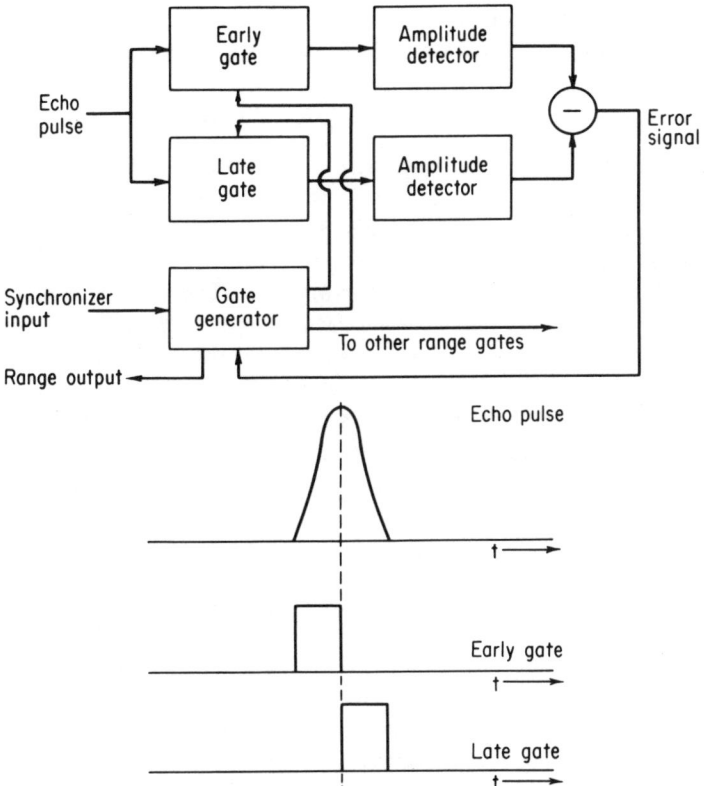

Figure 5-24. *Range tracking system.*

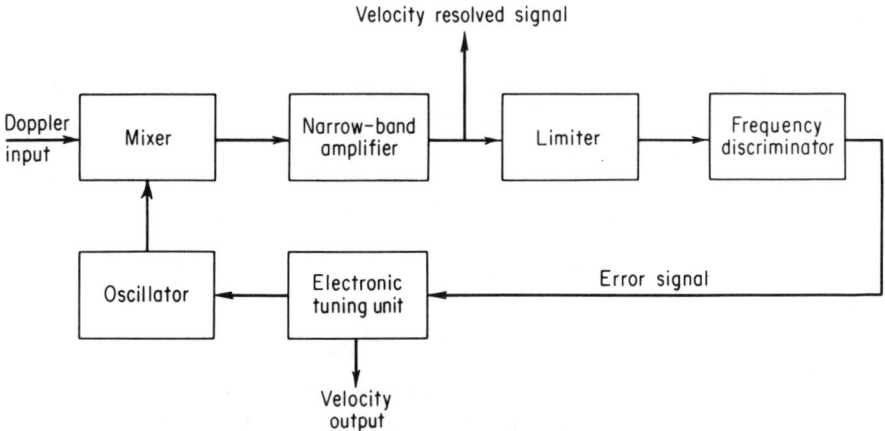

Figure 5-25. *Velocity tracking system.*

116 RADAR CONSIDERATIONS

in the radar, such as those which gate the *IF* echo pulses so as to provide range resolution in conical scan and *SLC* angle-tracking systems.

VELOCITY TRACKING

Radars that resolve in velocity, such as *CW* or pulse-Doppler radars, require some means of tracking any changes in the Doppler frequency. A velocity tracking system is shown in Fig. 5–25. The Doppler input signal is mixed with an oscillator signal to convert it to an intermediate frequency, which is then passed through a narrow-band *IF* amplifier. The frequency discriminator senses any difference between the *IF* and the discriminator center frequency, developing a voltage proportional to the error that tunes the oscillator, until the frequency difference is zero. The limiter removes any amplitude variations so that the error signal is independent of echo amplitude. The amount of tuning required supplies an indication of the velocity. The narrow-band amplifier provides the velocity resolution, and its output can be processed to obtain conical-scan modulation, FM ranging modulation, or other signals for which velocity resolution is desired.

GENERAL RADAR CONCEPTS

It is possible to consider the effects of various types of electronic countermeasures on each type of radar discussed in the preceding sections. However, since only specific radar systems have been discussed so far, the types of ECM would have to be equally specific: jammers for pulse radars, jammers for *CW* radars, and so on. The disadvantage of this approach is that it does not allow one to treat the effects of ECM on future radars (or classified radars not discussed here), and it misses several very fundamental concepts pertaining to radars and their ECM vulnerability. A better approach to the problem is to develop these fundamentals, relate them to the radar systems already presented, and then consider their implications in the problem of electronic warfare.

The detection range equations developed for pulse, *CW*, and pulse-Doppler (Eqs. 5–17, 5–30 and 5–40) radar systems are all identical. Considering the differences between pulse and *CW* concepts, they seem to be opposite approaches to the radar problem, and it is not intuitively obvious that one equation could describe the range of both systems.

The range equation is:

$$R^4 = \frac{\sigma A_e P_{av} T_s L}{(1.2) \, 4\pi \, kT \overline{NF} \psi_s \frac{S}{N}}. \tag{5-46}$$

The equation is placed in this form because the factors can be readily obtained from actual system specifications, simplifying numerical calculations. An even more basic form can be derived.

GENERAL RADAR CONCEPTS 117

When the radar is searching the solid angle ψ_s for the target, which is at range R, the power can be regarded as uniformly distributed over an area equal to $\psi_s R^2$, that portion of the spherical surface of radius R intersected by the antenna beam. Then the average power density at the target is:

$$\frac{P_{av}}{\psi_b R^2}. \qquad (5\text{--}47)$$

To obtain the average echo power density at the radar, Eq. 5–47 is multiplied by:

$$\frac{\sigma}{4\pi R^2}. \qquad (5\text{--}48)$$

This is the definition of radar cross section.

To obtain the receiving antenna waveguide power, the power density in front of the antenna is multiplied by A_e. A factor L is used to account for any losses. The echo energy, E, received from the target during one scan interval is T_i times the average power, or

$$E = \frac{P_{av}\sigma A_e L T_i}{\psi_b R^4 4\pi}. \qquad (5\text{--}49)$$

The range of the radar is limited by the noise from which the signal has to be extracted. In most cases this noise originates in the front end of the radar receiver, although low-noise solid-state devices such as masers and parametric amplifiers can reduce system noise to a level at which antenna noise or cosmic noise becomes the limiting factor. As in earlier discussions, the receiver noise is given by:

$$N_d = kT\overline{NF} \qquad (5\text{--}50)$$

where k is Boltzmann's constant, T is temperature, $\overline{NF}$ is the receiver noise figure, and N_d is the power density in watts per cycle per sec (which has the units of energy). Using Eqs. 5–16, 5–49, and 5–50, the ratio of signal power to noise power in the front part of the receiver is then:

$$\frac{S}{N} = \frac{E}{1.2 N_d}. \qquad (5\text{--}51)$$

As shown in Fig. 3–8, the probability of target detection is directly related to S/N, which is equal to the ratio of received target-echo *energy* to noise-power density. Thus, in theory, detection is independent of the type of radar modulation used, and pulse, *CW*, and pulse-Doppler radars are equally good. In practice such factors as clutter, target size, resolution and accuracy requirements, and ease of implementation can make one type of radar much more preferable, but they must all pay the same basic price for target detection: echo energy.

118   RADAR CONSIDERATIONS

Since a single detection-range equation applies for pulse, *CW* and pulse-Doppler radars, one would suspect that the same equation might hold for many other types of radars as well. It is possible to derive this equation for a general type of radar. Consider a radar that transmits and receives some arbitrary waveform. It can be pulsed, *CW*, or modulated in amplitude and frequency, as desired. Since the radar will be used to make successive measurements, its transmission will be assumed to be periodic, all periods identical, each period equal to T_p. Then the spectrum of the radar transmission will consist of a series of discrete lines, as shown in Fig. 5-26. Each

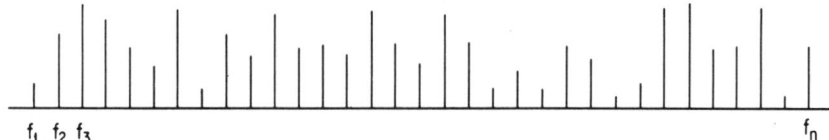

Figure 5-26. *Transmitted spectrum for a generalized radar.*

line is a sinusoid of constant amplitude and frequency and having some relative phase relationship to the other lines. Each is spaced apart in frequency by $1/T_p$. The sum of all these spectral components produces the waveform of the specified radar transmission:

$$\sqrt{2E_s}[a_1 \cos(\omega_1 t + \phi_1) + a_2 \cos(\omega_2 t + \phi_2) + \ldots a_n \cos(\omega_n t + \phi_n)]$$
$$= \sqrt{2E_s} \sum_{k=1}^{n} g_k(t). \quad (5\text{-}52)$$

It will be assumed that the target is stationary and isolated in space. The resulting echo will have the same form as Eq. 5-52, being different only because of round-trip propagation delay and amplitude reduction. Target detection requires making a decision, either "yes, there is a target," or "no, there is not a target." The echo must be processed so as to develop a voltage as far above the noise as possible. Since only a simple "yes–no" decision is required, this voltage should be continuous as long as the target is there and the antenna beam is on the target, such as d-c or a fixed sinusoid, rather than a pulse.

The highest voltage that can be obtained from the components given in Fig. 5-26 and Eq. 5-52 is simply the sum of all components, *added in phase*. This summation can be accomplished by first heterodyning all the components to a common frequency so that the phases between components will be independent of time. Next, each must be shifted in phase by $-\phi_1, -\phi_2, \ldots -\phi_n$ so they will all be exactly in phase. These two steps can be accomplished by simply multiplying or mixing each component by a signal having the same frequency and phase (and offset in frequency by an *IF* frequency if desired)

$$[E_s a_1 \cos(\omega_1 t + \phi_1)] \times [E_m b_1 \cos(\omega_1 t + \phi_1 + \omega_{if} t)]. \quad (5\text{-}53)$$

Filtering out the resulting difference frequency gives

$$\tfrac{1}{2}(E_s a_1)(E_m b_1) \cos(\omega_{if} t). \tag{5-54}$$

The phase shift, ϕ_1, is gone and ω_{if} is the desired common frequency. Thus, to create the largest possible sum from all the components in Fig. 5–26, these components should be multiplied by (i.e., mixed with) a second set of components that are identical in relative phase and frequency spacing to the first set. This reference set of components can differ in frequency in all its members from the other set by some Intermediate Frequency (*IF*), if desired. All the cross-product terms resulting from multiplying the series of Eq. 5–52 by a similar series are lost when the desired d-c component is filtered through a low-pass filter, or the desired *IF* signal is filtered through a narrow band-pass filter. This result is possible because all cross-product terms differ in frequency from the desired terms by $1/T_p$ or multiples of $1/T_p$.

The relative amplitudes of the signal components are given by $a_1, a_2, ..., a_n$ and those of the reference signal by $b_1, b_2, ..., b_n$ (Eq. 5–52), and as can be seen from Eq. 5–54, the b's can be regarded as amplitude "weighting factors" applied to each signal component. Obviously, the largest signal components in Fig. 5–26 should be weighted most heavily, whereas the smaller components that are further down in the receiver noise should be weighted less, and very small signal components lost in the noise should be multiplied by a weighting factor of nearly zero if the sum is to have the highest ratio of signal to noise. The amplitude of the sum of all the signal components is

$$S = \tfrac{1}{2} E_s E_m (a_1 b_1 + a_2 b_2 + ..., a_n b_n). \tag{5-55}$$

The amplitude of the noise voltage accompanying each signal component can be represented as

$$\tfrac{1}{2} E_m b_1 N_d, \quad \tfrac{1}{2} E_m b_2 N_d, \; ... \tag{5-56}$$

where N_d is the noise voltage (*RMS*) per cycle per sec accompanying the signal. Since the noise adds on a power basis, the resulting noise combined with the signal is:

$$N = \tfrac{1}{2} E_m N_d \sqrt{b_1^2 + b_2^2 ... b_n^2}. \tag{5-57}$$

Taking the signal-to-noise voltage ratio and maximizing S/N as a function of $b_1, b_2, b_3, ...$ gives the result that $a_k = b_k$. Thus, the echo and reference signal frequency components should both have the same relative amplitudes for maximum signal-to-noise ratio.

In addition, they are the same in frequency and phase, making them identical in all respects. Thus the optimum reference signal is a replica of the echo signal.

Multiplying the received signal by a reference signal and filtering out the resulting steady-state component is nothing more than one method of performing the correlation described by Eq. 3–10 in Chapter 3. The process was

120 RADAR CONSIDERATIONS

arrived at here by a consideration of the signal in the frequency domain with the goal of obtaining the largest possible sum from the signal components.

Processing the received signal by mixing it with an optimum reference signal does not alter the signal-to-noise ratio, as can be seen in Fig. 5-27. The input spectrum is arbitrary, corresponding to some general transmission (pulse, FM, phase-coded, and so on). After mixing, all the components are in phase and the output is a sine wave at the *IF* frequency (filtering removes

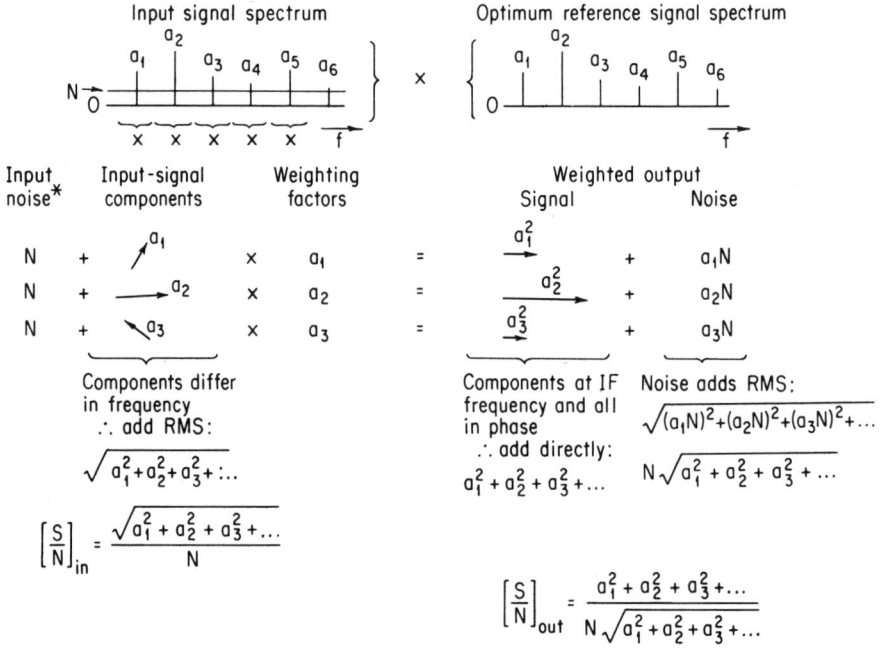

*Noise intervals "x" do not overlap in the input. In the output they are all heterodyned to a common center frequency and must be added.

Figure 5-27. *Optimum signal processing. Mixing the input signal with an optimum reference signal does not change the signal-to-noise ratio.*

all cross-product frequency components). It then remains for an operator or a decision circuit to detect the signal in the presence of the noise. (The values of S/N required for a given probability of detection are given in Chapter 3.) The derivation leading to the signal-to-noise ratio in Eq. 5-51 was made independent of the type of radar, and since this S/N is not changed by optimum processing, it follows that all radars can have a basic range equation.

The conclusion is that whatever form of modulation a radar uses, the echo signal should be processed by multiplying it by (mixing it with) a reference signal that has the same spectral components, and thus the same waveform, as the echo itself. Thus, all the spectral components are converted to

GENERAL RADAR CONCEPTS   121

the same frequency, arranged in phase, and added together to produce the maximum possible voltage. It would seem likely, then, that this sort of processing would be found in the pulse, *CW* and *PD* radars.

The *CW* radar is the easiest to evaluate. Its transmission spectrum and echo spectrum (assuming a fixed-point target) are a single line. The local oscillator supplies a coherent reference signal that is also a single spectral line, and the narrow-band Doppler filter is the decision circuit.

The pulse-Doppler radar echo is a series of *RF* pulses. The receiver multiplies the echo first by the coherent local oscillator and then later by the range-gate pulse. The same result would be obtained if the continuous local oscillator were pulsed, or gated on, by the range-gate pulse. Then this new local oscillator would have the same waveshape and spectral nature as the echo, meeting the requirements for an optimum reference signal. Once again the narrow-band Doppler filter serves as the decision filter and sorts out the sum voltage.

Note that the range-gate pulse must occur at the same instant as the return echo if the sum voltage is to be at its maximum. This relationship is obvious when the echo and reference waveforms are considered on a time basis, but not when their spectra are considered. The spectra are both continuous in time (during the interval when the beam is on target), and it is not obvious that a slight time delay between them can cause the resultant of their product to diminish to zero. The time delay, τ, introduced into Eq. 5–52 results in terms of the form

$$\cos\left[\omega_1(t + \tau) + \phi_1\right] \quad (5\text{–}58)$$

for the reference signal. This result is equivalent to a phase shift of $\omega_1\tau$ radians. Each of the terms in Eq. 5–52 will be shifted by a progressively increasing phase shift, the last term being shifted by $\omega_n\tau$. If mixing the echo and reference signals originally resulted in n vectors all in phase, the progressive time-delay phase shift results in the vectors opening into a fan, as shown in Fig. 5–28. When

$$\omega_n\tau - \omega_1\tau = 2\pi \quad (5\text{–}59)$$

or $\quad f_n - f_1 = \dfrac{1}{\tau}$

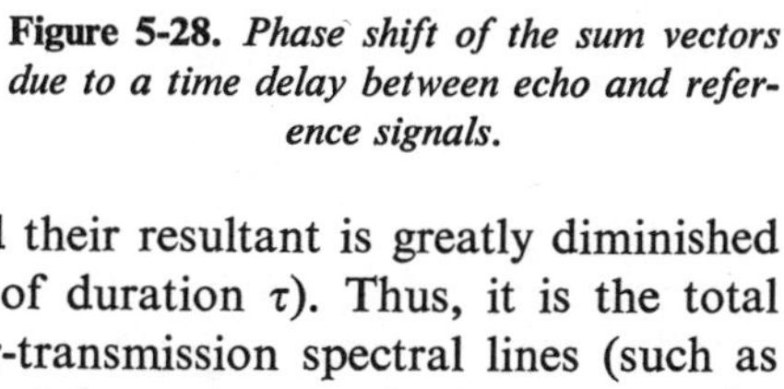

Figure 5-28. *Phase shift of the sum vectors due to a time delay between echo and reference signals.*

the vectors form a complete circle and their resultant is greatly diminished (to zero if the waveforms are pulses of duration τ). Thus, it is the total frequency range over which the radar-transmission spectral lines (such as those shown in Fig. 5–26) are spread that determines the basic range resolution of the radar.

122   RADAR CONSIDERATIONS

The pulse radar receives the same pulses as those received by pulse-Doppler. It also mixes the echo with a local oscillator. In the search mode the range gating is not readily apparent. For example, consider the radar display (range vs. angle) and visualize a miniature photocell observing one particular range on the display-tube face. It can be seen that this method of observation amounts to a form of range gating since the photocell can observe only echoes that correspond to a certain range. The eye and brain of the observer take the place of the photocell and perform this range gating. In this sense, the pulse-radar echo is also multiplied by a pulse reference signal, which meets the requirements for optimum echo signal processing.

It has been assumed that the generalized radar transmission, resulting in the spectrum in Fig. 5–26, is obtained by some form of modulation of a CW carrier. Another way to generate a variety of spectra is to pass a given spectrum (such as results from narrow pulse) through a filter. The filter transfer impedance will alter the input spectrum to some new spectral arrangement. These filters can be combined with various forms of modulation to produce a wide variety of different transmission spectra.

It has also been assumed that the optimum processing would be done by mixing the received signal with a replica reference signal, as was done in Eqs. 5–53 and 5–54 and illustrated in Fig. 5–27. If the spectral components in Fig. 5–26 are not all heterodyned to a common frequency but remain spaced apart by a frequency of $1/T_p$, the relative phases between components are continually changing with time. In this case, all the components can be arranged to be exactly in phase at some instant (such as $t = 0, T_p, 2T_p, ...,$) by passing the components through some special filter that has a phase response of $-\phi_1, -\phi_2, ..., -\phi_n$ at the frequency of each component. The filter aligns all the terms in Eq. 5–52 to the same phase. At this instant, the output of the filter will have the largest possible amplitude, resulting in a pulse. Consideration of the rate of change of phase between the complete sets of components shows that the duration of the pulse is equal to the reciprocal of the total frequency spread of the spectral components. This special filter is termed a "matched filter" and has the property of maximizing the ratio of peak pulse amplitude to RMS noise density. The matched filter must also have the correct "weighting factors" or transfer gain at each component frequency. If in Fig. 5–27 the "optimum reference-signal spectrum" is regarded as the filter transfer function, (a curve can be fitted to the tops of spectral lines a_1, a_2, a_3, and so on), and the output components are not "at IF frequency and all in phase" but are all in phase only momentarily at $t = 0, T_p, 2T_p, ...$, then Fig. 5–27 shows that the ratio peak signal power to the average noise power at the output terminals of the filter is equal to the ratio of average signal power to average noise power at the input. In this sense, correlation processing and matched-filter processing are the same.

THE EFFECTS OF ECM ON RADARS

In the preceding sections of this chapter we have examined three types of radars: pulse, CW and pulse-Doppler, and have postulated a generalized type of radar, examined its basis of operations, and related it to these three radars. This background now provides the basis for an examination of the effects of various types of ECM on radar systems.

The generalized radar transmission has the spectrum illustrated in Fig. 5–26, which consists of n spectral lines spaced uniformly in frequency by intervals of $1/T_p$ (because the transmission is periodic and each period is of duration T_p). The lines are arranged to have some arbitrary distribution of amplitudes and relative phases. In general, the enemy has no advance knowedge of how these amplitudes and phases are arranged (assuming security precautions deny him radar design details), so he cannot readily construct a jammer that *duplicates* the radar transmission. He can, however, *repeat* the radar signal, but the repetition may only accentuate the radar echo unless he can delay the echo as well so as to make the false repeated signal appear at a different range or angle.

It may be possible for the enemy to readily deduce the spectral arrangement of the radar transmission by simply observing its waveform. A plain pulse transmission is easily determined in spite of its extensive spectral composition. CW radars can also be detected from their transmission. But pulse and CW radars are only specific types in a large general class of possible radars.

Suppose the enemy does attempt to jam the generalized radar but fails to arrange the phase of the n spectral components in the same manner as the radar transmission. Then upon reception the radar receiver will process the jammer signal as it would a real target echo by mixing it with the proper reference signal. Since the jammer-signal spectral components are incorrect, the processing will not result in the summation of n in-phase voltages, but of n voltages having random phases, as illustrated in Fig. 5–29. If the amplitudes of the individual voltages are $a_1, a_2, \ldots, a_n$, for the desired echo, they add in-phase as:

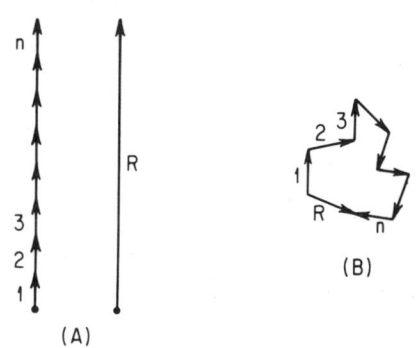

Figure 5-29. *The resultant, R, of n voltages added all in phase* (A) *and added with random phases* (B).

$$a_1 + a_2 + a_3 + \ldots, a_n. \tag{5-60}$$

For the random phases of the ECM signal they add as:

$$a_1^2 + a_2^2 + a_3^2 + \ldots, a_n^2. \tag{5-61}$$

If the relative amplitudes of the individual spectral components are all approximately equal, then the ratio of signal voltage to ECM voltage is:

$$\frac{na_1}{\sqrt{na_1^2}} = \sqrt{n} \qquad (5\text{--}62)$$

and the ratio of echo to ECM powers is n. Thus, the ECM is degraded in favor of the echo in proportion to the number of major spectral components in the echo. For example, the pulse-radar spectrum shown in Fig. 5–5 contains $1.2/\tau_o f_r$ spectral line within the band width Δf (assuming $\tau_o \Delta f = 1.2$). A jammer that attempted to produce pulses having this spectrum but was unable to get the proper phase arrangement (resulting in random phases) would be degraded by a factor of approximately $\tau_o f_r$, which is the ratio of pulse peak to average power.

It should be obvious that a *CW* radar, having only a single spectral line, cannot code its spectrum so as to to reduce the effects of a *CW* jammer. A pulse radar is not immune either, unless the phase arrangement of its spectrum is altered in some way so as not to be obvious to the enemy.

As an example of another radar with some ECM immunity, consider the *MTI* system shown in Fig. 5–19. Remove the delay line unit and substitute a range gate and bank of narrow-band Doppler filters, such as used in the pulse-Doppler system shown in Fig. 5–17. The *IF* amplifier and reference oscillator outputs are mixed to produce signals that are gated and filtered as shown in Fig. 5–18. The magnetron phase is different for each transmitted pulse; also, the *IF* reference oscillator is adjusted to be in phase with each transmitted pulse, so the *IF* echoes will be coherent. A jammer that generates radar-like pulses would also have to adjust the phase of each pulse, in keeping with magnetron pulse-to-pulse phase changes; otherwise the spectrum to each Doppler filter would be reduced in amplitude, as described in the section on *MTI* radars. Practical implementation of this hybrid radar is complicated by the pulse-Doppler requirement that the pulse-repetition rate exceed the expected Doppler, and the requirement for the circuit in Fig. 5–19 that the interval between pulses be greater than the round-trip propagation time.

Any coherent pulse radar, such as the pulse-Doppler system shown in Fig. 5–17, requires that the jammer pulses be coherent as well. Complete lack of coherence, such as would result if a conventional pulse jammer (one for non-coherent pulse radars) were used against a pulse-Doppler radar, would result in the jammer inefficiency described by Eq. 5–62.

The use of various types of coded radars, such as matched-filter or pulse-compression systems, can reduce the effectiveness of a jammer that attempts to duplicate the radar transmission by the factor described by Eq. 5–62. As the radars are designed to make them less vulnerable, it becomes more difficult to defeat them by duplication, and more attractive to simply "drown" the radar echoes by transmitting wide-band random noise. Consider the

EFFECTS OF ECM ON RADARS

jamming situation shown in Fig. 5-30, in which the radar is attempting to locate a target aircraft. In the absence of any jamming, the echo signal must be detected in the presence of receiver noise, and from Eq. 5-46:

$$\frac{S}{N} = \frac{P_{av}\sigma T_s L A_e}{(1.2)4\pi k T\overline{NF}\psi_s R_o^4}.$$

The target is detected at a range R_o.

When the barrage noise jammer is turned on at a distance R_j from the radar, the receiver noise-power density is given by:

$$\left(\frac{D_j G_j}{4\pi R_j^2}\right)\left(\frac{G_r \lambda^2}{4\pi}\right) + kT\overline{NF} \qquad (5\text{-}63)$$

instead of $kT\overline{NF}$. The jammer noise density (w/unit frequency) is D_j, gain of the jammer antenna in the radar direction is G_j, and the gain of the radar

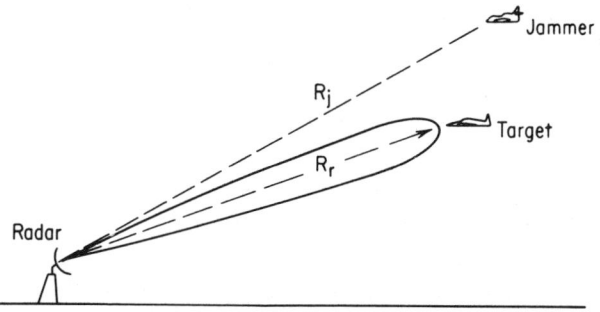

Figure 5-30. *Geometry of jamming and target aircraft.*

antenna in the direction of the jammer is G_r. Then the signal-to-jamming ratio is:

$$\frac{S}{J} = \frac{P_{av}\sigma T_s L A_e}{(1.2)4\pi\psi_s R_j^4\left(\dfrac{D_j G_j G_r \lambda^2}{(4\pi)^2 R_j^2} + kT\overline{NF}\right)} \qquad (5\text{-}64)$$

where the new distance from radar to target is R_j.

If the radar is to have the same signal-to-noise ratio for target detection in the presence of jamming as without,

$$\frac{S}{N} = \frac{S}{J}. \qquad (5\text{-}65)$$

Then the fractional reduction of radar-detection range due to noise-barrage jamming, R_j/R_o, is:

$$\left(\frac{R_j}{R_o}\right)^4 = \frac{(4\pi)^2 kT\overline{NF} R_j^2}{\lambda^2 D_j G_j G_r + (4\pi)^2 kT\overline{NF} R_j^2}. \qquad (5\text{-}66)$$

In most cases of interest the jammer noise is considerably greater than the receiver noise. If it is not, the jamming is not effective.

This large ratio of jammer noise to receiver noise allows simplification of Eq. 5–66 to:

$$\left(\frac{R_j}{R_o}\right)^4 = \frac{(4\pi)^2 kT\overline{NF}R_j^2}{\lambda^2 D_j G_j G_r}. \tag{5-67}$$

Whatever value of S/N existed at range R_o in the absence of jamming, will be recovered in the presence of noise jamming at range R_j. At any given range the ratio of S/N with and without jamming is also given by Eq. 5–67:

$$\frac{(S/N)_j}{(S/N)_o} = \frac{(4\pi)^2 kT\overline{NF}R_j^2}{\lambda^2 D_j G_j G_r}. \tag{5-68}$$

As an example, consider a radar operating at 10 $k$mc, having a side-lobe gain in the direction of the jammer of 10 db, and a receiver-noise figure of 10 db. A jammer at a range of 20 nautical miles radiates a noise-power density of 1 w/megacycle with an antenna gain of 10 db.
Then

$$\frac{(S/N)_j}{(S/N)_o} = 0.1$$

and

$$\frac{R_j}{R_o} = 0.56.$$

Thus the jamming reduces the signal-to-noise ratio by 10 db and reduces the single-scan detection range to 56% of the non-jamming range.

If the jammer is in the target, $R_j = R_r$ and G_r becomes the gain of the radar antenna main lobe, which simplifies Eq. 5–67 still further. Note that if the radar is expected to have increased detection range because of a lower noise figure, this range increase is more readily denied by jamming than if the range were obtained by increased radar power.

The result given in Eq. 5–67 is independent of the type of radar. The detection range in any radar is limited by the noise with which the echo signal must compete. The noise jammer simply increases this noise, or, in other words, it increases the effective receiver noise figure or temperature in Eq. 5–4 (also discussed in general terms in Chapter 3).

RADAR COUNTERMEASURES TECHNIQUES

As various types of radars are developed and become operational, they create a requirement for countermeasures to defeat any enemy radars of the same types. The decision as to kinds of ECM equipment that should be developed depends upon an assessment of the enemy's technology, intelligence data on his operating frequencies, and types of radars. In the absence of such intelligence, it is often assumed that the enemy has achieved a state of art comparable to ours; therefore his radars should be similar to ours. The problem is therefore reduced to devising ECM techniques to defeat our radar systems. It takes some time for a new radar technique to become sufficiently developed to show promise of eventually becoming operational, and more time for someone to decide to devise countermeasures. As a result, ECM development lags behind radar development. Likewise, radar counter-countermeasures are often prompted by new advances in ECM techniques.

One of the earliest types of ECM, and perhaps the simplest in theory, is noise jamming. A noisy signal is transmitted with sufficient power to completely obscure the target echo in the radar receiver, increasing the receiver noise level, as discussed in the preceding section. At lower radar frequencies the noise signal can be developed at low frequencies, then heterodyned to radar frequencies and amplified in power by conventional radio power tubes. The noise source can be a thermal diode, photomultiplier, or some other inherently wide-band noise device.

At higher radar frequencies CW microwave power sources, such as magnetrons or traveling wave tubes, can be modulated with noise or noiselike signals to spread their spectra so as to produce wide-band barrage jamming. Frequency modulation will spread the transmitted spectrum over a wide band, as shown in Fig. 5–14. The spacing between the resulting spectral lines is equal to the modulation frequency; any single line is CW. When several lines pass through the receiver band width they will add in-phase occasionally but out-of-phase most of the time, as illustrated by the vectors in Fig. 5–28. If many lines are received the effect is best interpreted as a single sinusoid, sweeping in frequency, which periodically passes through the receiver band width and is thus observed only a fraction of the time. Since such periodic disturbances are usually less damaging than continuous interference, the modulation frequency is arranged so that only a few spectral lines are within the receiver band-pass. Amplitude modulation by random noise can be combined with the FM to give each FM spectral component additional side-bands. Although FM plus random AM will spread a CW transmission over a wide band of frequencies, the result is not exactly equivalent to wide-band white noise, and special signal processing circuits can be added to the radar receiver that take advantage of the special nature of such jamming signals to reduce their effectiveness. However, any circuits that could reduce pure white noise barrage jamming would be equally effective against receiver noise. They would presumably be used to increase the basic range of the radar and thus would not be classified as an ECM "fix."

Considerable effort is spent in making radar and communications power-transmitting sources as stable and noise-free as possible, which often makes them unsatisfactory for barrage jamming. A more complete solution to the problem of generating barrage noise over a wide frequency range can sometimes be achieved by developing special microwave sources that are inherently noisy and unstable. For example, instead of using a *CW* magnetron and attempting to spread its spectrum over a wide band by modulating its power supply to induce frequency "pulling," it is possible to construct a magnetron that is basically very noisy. This expedient simplifies the overall design of the barrage jammer.

Since the enemy is uncertain as to the exact radar frequencies he must jam and the extent to which the radars can change frequency, he must be prepared to jam a wide band either by observing radar frequencies and rapidly tuning the jammers, or by covering a band sufficiently wide to counter all expected radars. Since a brief loss of jamming can allow the radars to assess the raid, the latter technique of wide-band jamming is often used in spite of its greater power requirement. Barrage jammers do not require the elaborate receivers, decision- or data-processing circuits, or automatic high-speed tuning units used in more sophisticated jammers, which improves their reliability and gives them some saving in power.

Noise jamming, either wide-band barrage or spot, serves to obscure the target return and prevent the radar from measuring its range or velocity. However, the radar can often determine the angular direction of the jammer. This is especially true for monopulse angle-sensing systems that are immune to signal modulations. Angle information will allow several radars to assess the target range by triangulation. Noise jamming can even call attention to the attack and be used as a "beacon" to guide attacking aircraft. More subtle approaches to jamming are often necessary.

The lowest-power jammer possible is one that duplicates the radar signal, thus taking full advantage of the radar sensitivity. It is easy to duplicate the pulses of a non-coherent pulse radar since only the pulse waveform and frequency are important. Once the jammer is on the correct frequency it can transmit pulses at any desired time, without regard to the phase of the radio-frequency carrier. A false target can be made to appear at a range greater than that of the target by simply transmitting a pulse a short time after the jammer receives the pulse from the radar (12 μ sec of delay for each nautical mile). It is more difficult to make a target appear to be at a closer range since the jammer must anticipate the arrival of the radar pulse by a short time interval. To do so, the jammer must make use of the radar pulse-repetition rate. A varying pulse rate can trick the jammer into making mistakes and reveal the presence of the false target.

To generate false targets at various angles, the jammer determines the rotation rate of the radar antenna and responds whenever the antenna is pointing in the desired direction. A constant rate of rotation is assumed. The reduced antenna gain on either side of the main beam is compensated for by transmitting stronger pulses. Thus, at longer ranges the angular span within which false targets can be generated is less than at shorter ranges, because of the maximum-power limitations of the jammer. Any variations in pulse timing will cause the false target to fluctuate in position and reveal its identity. Careful jammer design can allow one aircraft with a deceptive jamming repeater to simulate an attack by many aircraft, complete with programmed maneuvers for each aircraft.

The generation of false targets can be greatly simplified if they are not required to fly consistent courses. A random generation of many false targets can be very confusing, since the operator must eliminate false targets by determining that they do not appear on two or more successive scans. When the radar rotates 360 degrees for each scan, the detection of real targets among a screen of random false targets can be a lengthy process. Once again, radar-antenna directivity and jammer power limitations will restrict the angular sector that can be jammed. At short ranges, however, a full 360 degrees of the display can usually be covered.

Coherent radar systems require that the repeater-jammer maintain the same frequency and phase stability employed by the radar transmitter; otherwise the jammer's signal will not be properly processed, resulting in decreased jamming efficiency and possible indentification of the false targets. For radars that use pulse compression, pulse coding, and matched-filter techniques, the jammer cannot readily duplicate the complex radar transmission. The most feasible technique is for the jammer to repeat the signal exactly as it receives it. This technique requires some sort of signal-delay device that has sufficient band width to pass the radar modulation. An adjustable delay is needed if the real and false targets are to have relative motions. Such delays are difficult to obtain, especially those long enough to generate false targets at ranges shorter than that of the jamming aircraft. Coding the radar signal so that it covers a wider frequency band complicates the delay problem still further.

Some tricks can be used to partially defeat false target repeaters. An angle-tracking antenna can be used in conjunction with the normal search radar to determine the directions of the repeater jammers. The minimum ranges to the jammers can also be obtained by shifting the radar pulse rate.

Once a radar has located the desired target and has commenced tracking, generation of false targets is no longer useful. Tracking also assumes that any noise jamming has been overcome. Special jamming techniques can be used to unlock the tracking radar or degrade tracking accuracy.

The first step is to generate a false echo and superimpose it upon the real echo. Then, by making the false echo stronger, the radar is caused to track the false echo instead of the real one.

Conical-scan angle-tracking systems can be defeated by sensing the scan-rotation frequency and amplitude modulating the false echo at this frequency. As described in the section on angle tracking, the scan-modulation signal is used by the radar to direct the antenna servos. The false modulation forces the antenna away from the correct target direction. The amount and direction of offset depends upon the amplitude of the modulation and the relationship of its phase to the scan orientation of the beam. The effects of inverse conical-scan jamming can be defeated by using monopulse angle-tracking systems.

The large false target can be used to interfere with the operation of range and velocity-tracking systems. Range tracking can be broken by gradually moving the false target out in range, thus pulling the range gate away from the true target, and then turning off the false target. The radar then has to reacquire the true target within the range gate.

Velocity track breaking is accomplished by gradually shifting the Doppler of the false target so as to move the Doppler tracking gate away from the true target Doppler, and then turning off the false target. Although target-range or velocity tracking can usually be re-established (unless the radar is automatic and has no relock provisions), the disturbance caused by continual loss of lock can degrade the range or velocity information sufficiently to defeat any attack against the jamming aircraft. It also reduces the target-handling capabilities of automatic track-while-scan radar systems in which one radar simultaneously tracks many targets once the operator has established initial target lock-on. Track breaking would keep the operator busy re-acquiring unlocked targets.

A commonly used countermeasure technique is to drop chaff from the jamming aircraft. Chaff consists of small metallic dipoles that are designed to resonate at the radar frequency and thus have a very large radar backscatter cross section for their weight. Thousands of dipoles are compressed into small packages that burst open when tossed into the aircraft slipstream, scattering the dipoles into a cloud.

The overall radar cross section of a chaff cloud of N particles is approximately

$$\sigma = 0.18\lambda^2 N,$$

where λ is radar wavelength, and the dipoles are assumed to be uniformly oriented in all possible directions. If the particles are $\lambda/2$ long, 0.01 inches wide and cut from aluminium foil 0.001 inches thick, the cross section is

$$\sigma = 30,000 \frac{W}{f},$$

where σ is in sq ft, W is total chaff weight in pounds, and f is radar frequency in kmcs. For example, 0.1 pound of chaff cut for 10 kmcs would have a cross section of approximately 300 sq ft. Since the dipoles are resonant, they cover only a limited frequency range, roughly $\pm 10\%$ to $\pm 15\%$, and must be cut to different lengths to cover wider bands. Assuming $\pm 10\%$, 0.5 pound of chaff could give a 300 sq ft cross section for all frequencies between 1 kmc and 10 kmcs.

Each chaff package, dropped independently, can simulate an additional aircraft. A chaff curtain dropped by a forward line of aircraft, consisting of thousands of false targets, can so confuse the radars that they are unable to locate the real targets within the chaff area. Chaff drops so slowly that it can take days to reach the ground.

If the chaff packages are dropped in close sequence, the resulting continuous chaff corridor has so great an angular extent when viewed at right angles that angular tracking systems have difficulting establishing a definite target to track. As a result, the tracking often wanders off the target. If the corridor is viewed nearer to head-on (aircraft in front), the range gate can be forced to stay on the first return received, which will be from the aircraft, thus gating out the chaff echoes. When viewing the corridor tail-on the range gate is forced to select the last part of the echo, which is once again the aircraft. If, in addition to dropping chaff, rockets are used to fire chaff in front of the aircraft, the problem of maintaining tracking is greatly increased.

Since the chaff particles have high aerodynamic drag, their forward velocity quickly drops to zero while the aircraft continues to move. Thus, chaff can be regarded as an airborne type of "clutter." Radars such as CW and pulse-Doppler that can reject clutter are not seriously affected by chaff.

6

THE ROLE OF ANTENNAS IN ELECTRONIC WARFARE

The antenna is the transducer that intercepts electromagnetic waves propagating through the atmosphere and directs them into guided waves in a transmission line; or conversely, in the case of a transmitting antenna, it converts guided electromagnetic energy in a transmission line into unguided or free-space electromagnetic waves. In short, the antenna couples the system to the environment. The physical shapes and dimensions of antennas are extremely varied, depending upon their frequency, band width, power-handling capacity, polarization, and radiation pattern. They may range in physical size from fractions of an inch to miles in length, may handle frequencies from a few cycles to thousands of megacycles per second, radiate powers from microwatts to megawatts, and cost from a few dollars to tens of millions. Since every electronic-warfare system that either radiates or receives electromagnetic energy from the atmosphere must involve an antenna, it is evident that a discussion of antennas and their characteristics, is indeed necessary.

Much of antenna theory and practice even today is more art than science. To develop detailed design formulas for any particular antenna or antenna type is clearly outside the intent of this book. It is intended rather to provide the reader with an appreciation of the general principles of antenna performance and to relate these principles to the problems of electronic-warfare systems.

Similarly, the sections on propagation will not do more than introduce the reader to the basic principles of electromagnetic propagation. It is important to recognize that both antenna and propagation theory are tremendous fields in themselves. To restrict this chapter to subjects in these fields most closely related to electronic warfare it has been necessary to avoid

THE ROLE OF ANTENNAS   133

all but the most superficial analyses. In a typical situation the information provided by this chapter will permit the system analyst or designer to determine sufficient antenna parameters to provide antenna specialists with the information required to develop detailed design characteristics. If this goal is reached, the chapter will have served its purpose.

ANTENNA DESCRIPTIVE PARAMETERS

The first important characteristic of antennas that will be discussed is the radiation pattern or antenna pattern. The pattern of a transmitting antenna is a plot of the relative amplitude of the energy radiated as a function of angle about the antenna. Imagine a transmitting antenna located at the origin of coordinates of Fig. 6–1, and that a small pickup probe is moved in a circle around the antenna in the X and Y plane. If the signal received by the probe is plotted as a function of ϕ, a pattern similar to Fig. 6–2 might result. This would be described as a horizontal-plane pattern of the antenna. Similar data could be recorded by moving the pickup probe in a vertical plane (ϕ = constant) and plotting the received signal as a function of θ. This configuration would be known as a vertical-plane pattern. If the value of ϕ was such that the vertical plane intersected the nose of the beam it would be known as the principal-plane vertical pattern.

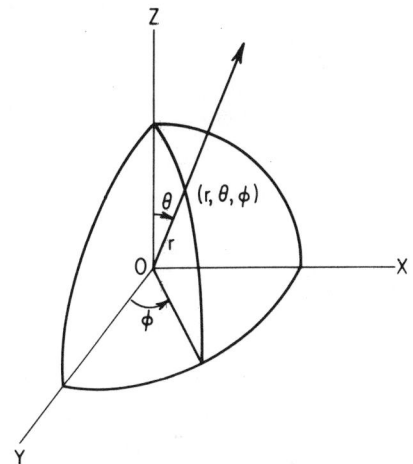

Figure 6-1. *Spherical coordinate system for antenna located at point 0.*

As a matter of convenience most antenna patterns are not recorded in this manner; instead, the measuring probe or antenna is held fixed and the antenna under test is rotated. Furthermore, the direction of propagation is usually reversed. That is, the antenna under test acts as the receiving antenna and the signal is radiated from the fixed antenna.

At this point it is necessary to refer to an important theorem of antenna theory, the "Reciprocity Principle," which may be stated as follows:

> If a voltage V applied at a point P in antenna A causes a current I to flow at a point Q in antenna B, then the same voltage V applied to antenna B at point Q will cause the same current I to flow at point P in antenna A.

From this theorem it can be shown that the transmitting and receiving patterns of an antenna are identical. A rigorous proof of the reciprocity

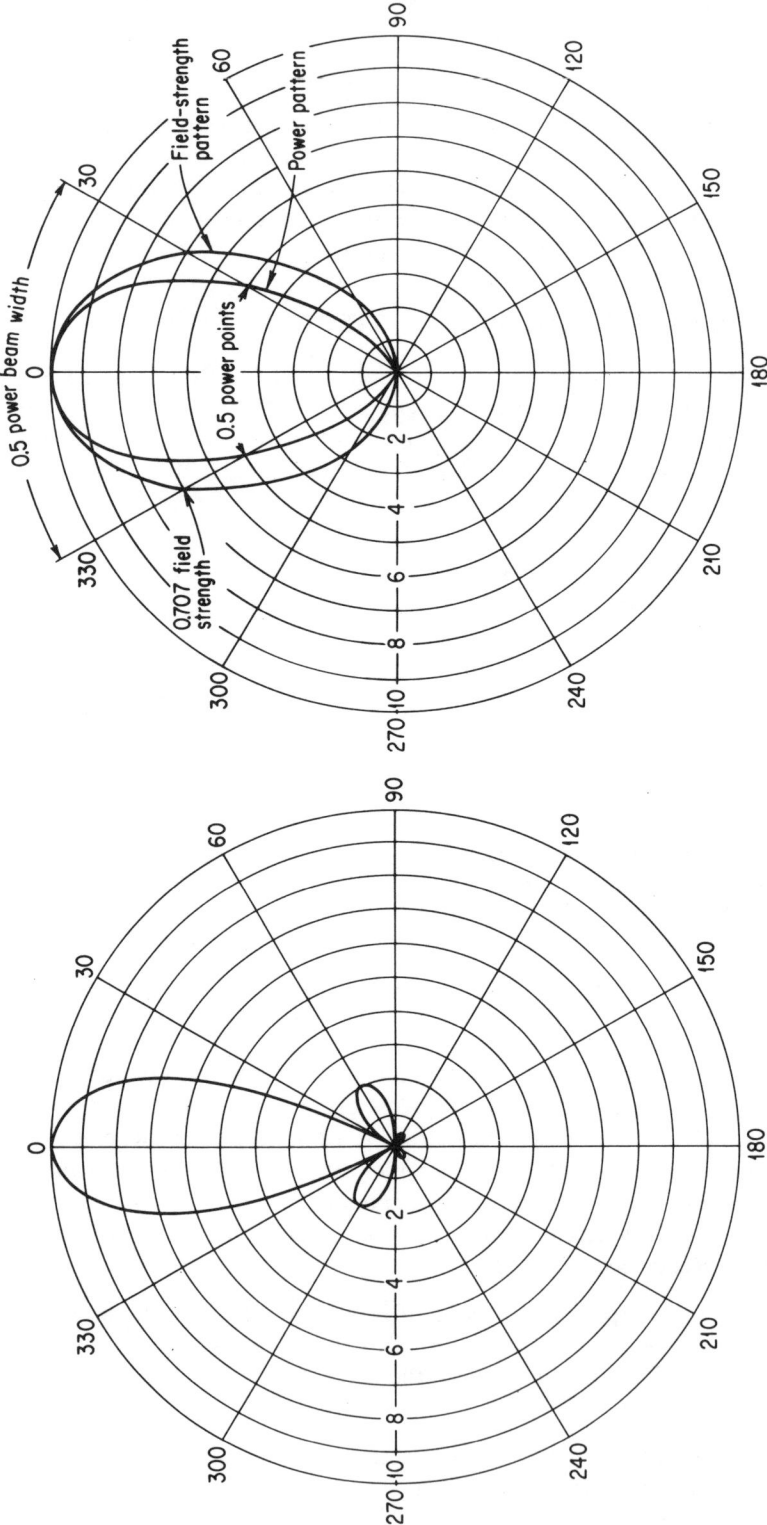

Figure 6-2. Horizontal plane pattern (XY, plane) resulting from moving a pickup probe about point 0.

Figure 6-3. Comparative plots of electric field strength and power density per unit area for the same antenna.

theorem can be found in Silver, *Microwave Antenna Theory and Design*,* but it will be sufficient here to state that for all antennas considered in this book the transmitting and receiving patterns are the same.

The pattern of an antenna may be plotted by recording the relative electric-field strength as a function of angle, or by recording the relative value of power per unit area as a function of angle. The former is known as a field-strength pattern, and the latter as a power pattern. Figure 6-3 is a superposition of the two plots from the same antenna, and it will be noted that the power pattern may be obtained by merely squaring each point of the field-strength pattern.

Highly directional antennas such as those used for radar, for example, are usually recorded as a power plot in rectangular coordinates (Fig. 6-4) with the abscissa representing the space angle and the ordinate a logarithmic plot of the relative power per unit area.

The main lobe of a directional antenna is usually defined by its half-power beam width, the angle subtended by the points where the power per unit area has dropped to one-half the maximum value, or in the case of a field-strength pattern, the point at which the field strength has dropped to 0.707 of the maximum. It should be noted that in all cases radiation patterns are for a single polarization and antenna parameters are usually discussed in terms of the principal polarization.

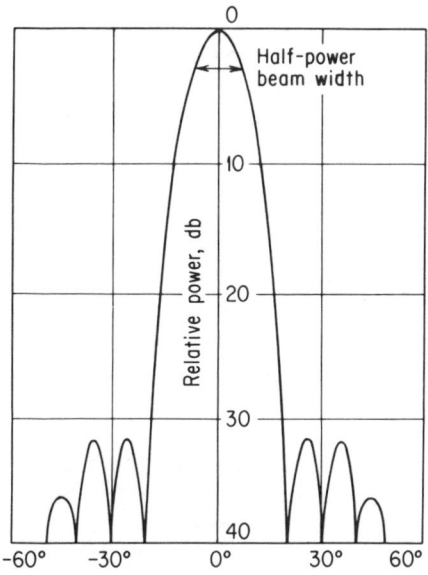

Figure 6-4. *Antenna power pattern. Relative power in db vs. azimuth angle.*

Polarization for an electromagnetic wave is the direction in which the electric-field vector is oriented, and the polarization of an antenna is the polarization of the electromagnetic field that would be radiated by the antenna. If the electric-field vector is fixed in a plane normal to the direction of propagation it is said to be linear or plane polarized. In the most general case however, the electric-field vector will describe an ellipse in the plane normal to the direction of propagation. This is known as right-hand elliptical polarization if the vector rotates in a clockwise direction when viewed from behind, and as left-hand elliptical polarization if it rotates in a counterclockwise direction when viewed from behind. Linear and circular polariza-

* Vol. 12, MIT Radiation Laboratory Series, McGraw-Hill Book Company, Inc., New York, 1949.

tions are obviously the limiting cases of elliptical polarization, depending upon the relative amplitudes of the semi-major axes of the ellipse.

DIRECTIVITY AND GAIN

The degree to which any particular antenna pattern is concentrated into a beam is known as "directivity", and may be defined as follows:

$$D = \frac{P_{max}}{P_{ave}}, \qquad (6\text{--}1)$$

where P_{max} = maximum power flux radiated and P_{ave} = average power flux radiated. Since the average power flux radiated is equal to the total power radiated (P_{tot}) divided by 4π steradians,

$$D = 4\pi \frac{P_{max}}{P_{tot}}. \qquad (6\text{--}2)$$

Directivity as defined above is a function of the antenna pattern only, and does not take into account the antenna efficiency or losses. In order to compare the relative performance of antennas in any electronic-warfare system, it is necessary to include the effect of the antenna efficiency. This effect is known as gain, G, defined as follows:

$$G = KD, \qquad (6\text{--}3)$$

where K is the efficiency factor ($0 \leq K \leq 1$). Thus it is seen that for a 100% efficient antenna ($K = 1$), the gain would be equal to the directivity; but for all practical antennas ($K < 1$), gain is less than the directivity. Gain is frequently expressed as a decibel ratio:

$$\text{db gain} = 10 \log_{10} G. \qquad (6\text{--}4)$$

A very important relationship exists between the gain and the physical antenna dimensions, and electrical wavelength:

$$G = 4\pi K \frac{A}{\lambda^2}, \qquad (6\text{--}5)$$

where: A is the area of the antenna, λ is the wavelength (both expressed in the same units), and K is the efficiency factor.

Practical values of K for directive microwave antennas lie between 0.5 and 0.8 depending upon the band width and side-lobe level, as will be discussed in the next section. For parabolic antenna systems the beam width between half-power points may be approximated by:

$$\theta = 70 \frac{\lambda}{D}, \qquad (6\text{--}6)$$

where D is the antenna diameter.

Another useful relationship relates the gain to the antenna beam widths:

$$G = K\frac{40{,}000}{\theta_H \theta_V} \tag{6-7}$$

where θ_H is the horizontal half-power beam width, and θ_V is the vertical half-power beam width (both expressed in degrees).

ANTENNA SIDE LOBES

It will be noted that even in highly directive antenna patterns all of the energy is not concentrated into the main beam. There is always radiation in directions other than the preferred one. These side lobes, as they are called, are of particular importance in electronic warfare, since they represent a major limitation to the system's effectiveness. Jamming signals or false targets received on a side lobe can cause confusion and failure of a mission far more easily than false signals detected by the main beam. The reduction and control of the position and magnitude of side lobes in one of the most difficult facets of the antenna designer's task, and is extremely important when attempting to localize target azimuth in a hostile electronic environment. Certain elements of the subject must be made familiar to the system designer in order that he may understand the limitations placed upon him by this side-lobe problem and that he may make reasonable compromises between required antenna gain, size, and side-lobe levels.

It is sometimes believed that if the total power radiated remains constant and the side-lobe levels are reduced, then the gain should increase because more power will go into the main lobe. It is true that more power does go into the main lobe, but the beam width of the lobe increases simultaneously so that the resultant gain at the beam maximum is reduced. Thus, it is an unfortunate fact that the reduction of side lobes for any particular antenna is accomplished at the expense of gain in the main beam. A uniform illumination across the apertures of antennas of the reflector type or a uniform current distribution in an array will result in the narrowest beam and maximum gain, but the side-lobe level will be quite high, only about 13 db below the level of the main beam. In order to reduce the side-lobe level it is necessary to taper the illumination across the aperture. By proper control of the illumination taper it is possible to reduce the maximum side-lobe level to 30 or 40 db below the peak, but the result is that the antenna aperture will not be as effective and the gain will be reduced.

BAND WIDTH

An important factor in limiting overall performance is the band width of the antenna—the range of frequencies over which the antenna is designed to operate. This range is usually expressed as percentage of band width with respect to the center frequency. It is much less difficult to design an antenna to operate at a specified fixed frequency than it is to cover a wide band of frequencies. Radar antennas are considered to be quite broad band if they will perform satisfactorily over a ± 5 per cent band, whereas ECM reconnaissance antennas may be required to cover an octave or more. In the latter case it will be noted from Eq. 6–5 that the gain will not remain constant. For a given antenna area, if the operating wavelength is to vary as much as two to one, then the gain will vary by a factor of four. In other words, the gain at the high-frequency end of the band will be approximately 6 db more than at the low-frequency end. The relationship is approximate because the efficiency factor will not remain constant either, but for most planning purposes it will be sufficient to remember that a two-to-one change in frequency will result in a 6 db change in gain.

In addition to the limitations placed on antenna systems by allowable variations of gain, beam width, side-lobe level and band width, there is the effect of impedance matching to be considered. Since the antenna is a coupling device between the transmission line and free space, it must be properly matched to the transmission line if it is to transmit all the incident power without reflection. An improperly matched transmitting antenna will reflect part of the power back down the line toward the transmitter with the resultant creation of a standing wave of voltage along the line. There are three serious consequences of a high Voltage Standing Wave Ratio ($VSWR$). The first, of course, is a loss of efficiency, since all the power is not radiated. The second is caused by the fact that most transmitters will not operate properly with a mismatched load. Their frequency of operation may become unstable, and their power output may drop off. This reaction of transmitting tubes to changes in load impedance is known as "pulling." The third and last result of a high $VSWR$ is a reduction in the power-handling capacity of the antenna and transmission line because of the high-voltage points that occur every half-wavelength along the line. The higher the $VSWR$ the greater the magnitude of the high-voltage points for a given power input to the line. Voltage breakdown will occur at one or more of the high-voltage points at a power level far below the power-handling capacity of the antenna and transmission line, if they were properly matched.

Receiving antennas are not as critically dependent upon the $VSWR$ as are transmitting antennas, since they are not required to handle high power levels. Thus the requirements for impedance matching of receiving antennas are not as rigorous as for transmitting antennas. Since any mismatch whatever does represent a loss because of power reflection, it is important to consider how much loss can be tolerated in a given system. If the $VSWR$ is represented as γ, then the percentage of power reflected by the impedance mis-match is given by:

$$\text{Percentage of power reflected} = 100\left(\frac{\gamma - 1}{\gamma + 1}\right)^2. \qquad (6-8)$$

From the explanation above it can be seen that an impedance mismatch with a *VSWR* of 6 will reflect 51 % of the incident power, or a loss of approximately 3 db. A curve of *VSWR* vs. percentage of power reflected is plotted in Fig. 6–5.

PROPAGATION

The behavior of electromagnetic waves as they traverse the atmosphere must also be taken into account. The problem includes not only an understanding of the mechanism involved in the radiation of energy at the transmitting antenna and the reception of energy at the receiving antenna, but also the

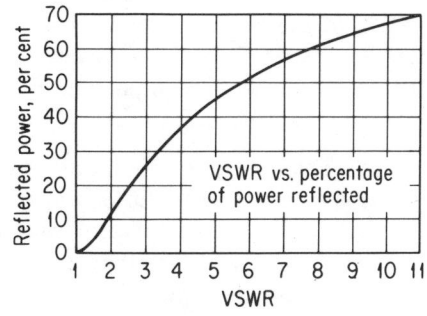

Figure 6-5

manner in which the atmosphere reacts upon the wave energy as it passes through. This topic is so huge and so difficult that many of the problems are neither completely understood, nor yet subject to exact mathematical analysis. Since the boundary conditions of the problem involve the surface of the earth on the one hand and the atmosphere on the other, it is obvious that exact mathematical descriptions of the phenomena involved in propagating electromagnetic energy through such a medium will be extremely complex. There are, however, a few general features of atmospheric propagation that should be familiar to the student of electronic warfare.

PROPAGATION OVER A PLAIN EARTH

Assume there is a transmitting antenna located at point A in Fig. 6–6, and a receiving antenna at point B (the earth in this case is assumed to be flat). The

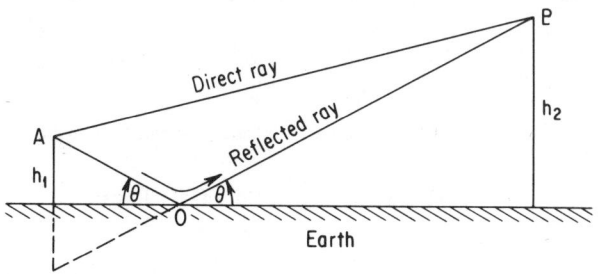

Figure 6-6. *Multipath condition for signal arriving at point B from transmitter located at A.*

signal strength at point B will result from the direct radiation from the transmitting antenna along the line AB plus a contribution to the field from the reflected ray traversing the path AOB. The phase of the reflected ray can

140 THE ROLE OF ANTENNAS

vary from 0 to 180 degrees, and its magnitude can range from zero to a value equal to that of the direct ray, depending upon the polarization of the signal and upon the reflection coefficient of the earth at the point of reflection. As an example of the extreme complexity of the problem it can be pointed out here that the reflection coefficient of the earth is a function of the terrain (sea, grassland, forests, and so on), the frequency, the angle of incidence, and the polarization. The important conclusion to draw from all the above discussion however, is that:

> The field strength at any point B due to a signal radiated at point A above a reflecting surface may vary from zero to twice the value that would exist if the reflector were not there (free-space value). The maximum values occur when the reflected wave arrives in phase with the direct ray, and the minimum values occur when the reflected ray arrives 180 degrees out of phase with the direct ray.

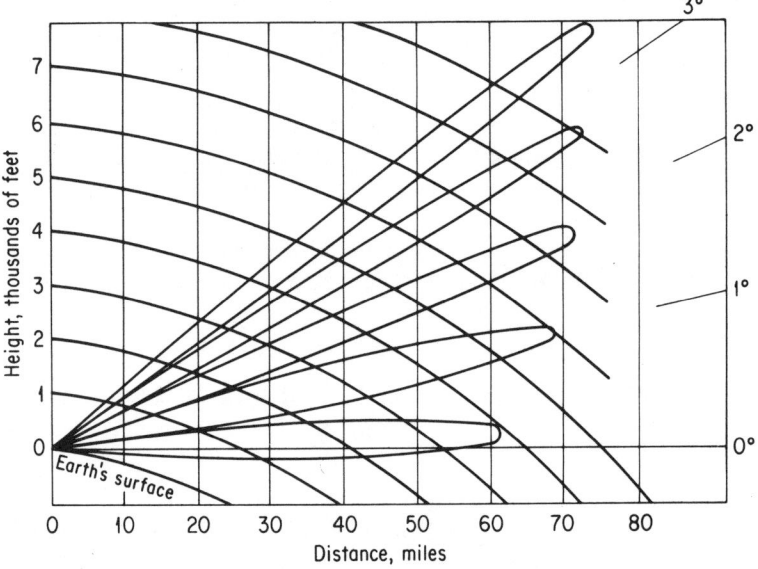

Figure 6-7. *Vertical coverage diagram of a radar antenna over a curved earth.*

PROPAGATION OVER THE CURVED EARTH

The effect of the curvature of the Earth is merely to complicate the geometry of the problem, but not the conclusion stated above. The vertical-coverage diagram will appear similar to Fig. 6–7. The number of lobes is dependent upon the height of the antenna above the earth in wavelengths, whereas the depths of the nulls and the magnitude of the lobes are dependent upon the reflection coefficient of the earth.

It is interesting to note that during World War II the vertical-lobe pattern of air-search radars was used for making a quick estimate of the altitude of aircraft detected by the radar. Since the initial detection of incoming raids was usually in the first lobe, noting the range of first detection and knowing the lobe angle gave an immediate first estimate of altitude. Noting the range where the targets first faded out and again when they reappeared enabled the radar operators to confirm or re-estimate the altitude of the raid. The knowledge and understanding of these nulls is of vital importance in electronic warfare. Examination of Fig. 6-7 will indicate how vulnerable a ground-based radar is to low-flying attackers. A plane flying a few hundred feet from the ground can approach quite near to the radar before entering the first lobe of the radar antenna pattern where it might be detected. Even an aircraft flying at a 1000 ft altitude can approach to within 20 or 30 miles of the radar before detection. Jamming and reconnaissance antennas are subject to the same kind of lobe structure when located near the surface of the earth. Consideration must be given to the location of the nulls in the vertical-coverage pattern to avoid surprise and insure proper antenna performance.

PROPAGATION IN THE STANDARD ATMOSPHERE

The region of the atmosphere extending from the surface of the earth to about 45,000 ft is known as the troposphere, and is characterized by the gradual decrease of temperature with altitude to a minimum of about -60 degrees C. Above the troposphere is the stratosphere, with temperatures either constant or increasing with altitude. At various altitudes above 200,000 ft are found layers of ionized gas known as the ionosphere. These layers act as reflecting surfaces for frequencies below 30 megacycles and play an important role in long-distance communications at low frequencies. At higher frequencies the waves penetrate the ionosphere and continue on into space. For this reason, it can be seen that above 30 megacycles long-range trans-horizon propagation will not be possible in the standard atmosphere. Anomalous atmospheric conditions, which allow the higher frequencies to propagate around the curvature of the Earth, will be discussed in the next section.

A so-called "standard atmosphere" has been defined, which approximates the most common tropospheric conditions in the Temperate Zone.

STANDARD ATMOSPHERE

Sea level		*Rate of change with altitude*
Temperature	15° C	$-6.5°$ C per km
Dry-air pressure	1013.2 millibars	-34 millibars per 1000 ft
Water-vapor pressure	10 millibars	-1 millibar per 1000 ft

The conditions defined above result in a medium with an index of refraction n, which decreases slowly with increasing altitude. The velocity, v, of an electromagnetic wave traveling through a refractive medium is equal to the velocity in a vacuum C, divided by n, the index of refraction. Since n is smaller at higher altitudes, the upper portions of a wave front will travel faster than the lower portions and the wave front will be refracted downward. This effect results in a slight tendency for radio waves to bend around the earth and to propagate over the horizon. In practice it has been found that this effect is equivalent to assuming that the earth has a radius equal to $\frac{4}{3}$ of its true radius, and that radio waves travel in straight lines. This assumption results in a simple and easily remembered formula for calculating the distance to the radio horizon. If the antenna is located at a height h ft above the earth, then the distance, d, in statute miles to the radio horizon is:

$$d = \sqrt{2h}. \tag{6-9}$$

Electromagnetic propagation within the line-of-sight region is characterized by attenuation proportional to the square of the distance plus an interference effect due to reflections from the earth as described in the previous section.

At distances greater than line of sight the phenomenon of diffraction becomes the dominant effect and the field strength falls off with a nearly

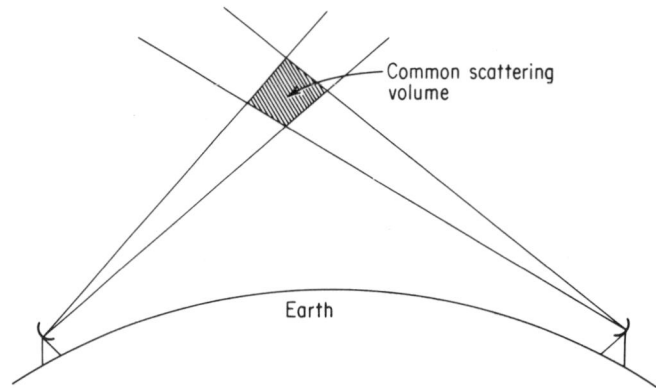

Figure 6-8. *Tropospheric scatter. Communication is possible beyond the line of sight due to scattering from a common volume subtending the beams from two antennas.*

constant rate per mile, as would be predicted from the theory of diffraction around a smooth sphere.

At great distances beyond the radio horizon it has been found, however, that the signals are much stronger than would be predicted by diffraction theory, and indeed the rate of change with distance is also much lower than in the diffraction region. Although the mechanism is not thoroughly understood, it has been suggested that some form of scattering occurs in the troposphere, resulting in greatly enhanced signals when two antennas are "looking" at a common volume as shown in Fig. 6–8.

Whatever the scattering may be, it results in fields that attenuate at a rate of about 12 db/100 miles. Very high-gain antennas have been built in order to extend the maximum range of tropospheric-scatter communication links. The same phenomenon makes possible the surveillance and intercept of electromagnetic radiation by means of very large, high-gain antennas at ranges far beyond the line of sight. It was found that the effective gain increased in direct ratio to the area as would be expected, up to about 40 db. Above this figure the gain does not increase as rapidly with increased antenna size as one would expect it to, based on free-space antenna theory. Instead, the gain increases roughly in proportion to the square root of the area. This degradation of antenna gain has been predicted theoretically and observed experimentally, but much more work is necessary to fully explain the phenomenon.

PROPAGATION IN A NON-STANDARD ATMOSPHERE

The standard atmospheric conditions that result in a gradually decreasing index of refraction with increasing height are frequently replaced by non-standard or anomalous conditions, which may drastically affect the propagation of radio waves. These anomalous conditions are known as ducts, and their effect upon propagation is known as trapping or ducting. A sudden increase of temperature at altitude caused by a warm-air mass overlying cooler air, such as might occur on a clear night when the earth cools rapidly, or when warm dry air from the land blows out over a large body of water, is called a temperature inversion. Radio waves are refracted more strongly downward by such inversions and may travel for greater distances around the curvature of the earth as a result. Such trapping does not often occur at frequencies below 200 megacycles but is quite common at microwave frequencies. The lowest frequency capable of being trapped by any given duct is a function of the thickness of the duct. Hence, strong inversions at higher altitudes will trap a larger band of frequencies than the thinner oceanic ducts, which are usually 100 ft or less above the surface.

In order for radiation to be trapped by a duct it is necessary that the antenna be located in or below the duct. Hence, there will be little or no effect upon propagation from aircraft antennas. Ground-based systems may experience greatly enhanced range performance due to the presence of ducts however, and in those regions of the world in which ducts occur frequently a meteorological analysis should be made to select the optimum height of antennas. The range of point-to-point UHF communications can therefore be extended and a certain degree of security and freedom from jamming will be gained.

ATTENUATION DUE TO ABSORPTION AND SCATTERING

At frequencies below 6000 megacycles, the signal strength varies with distance according to the familiar inverse square law. As the frequency increases and the wavelength becomes shorter, the loss due to molecular absorption by water-vapor and oxygen molecules becomes predominant. At a wavelength of 1.35 cm water-vapor absorption reaches a peak value of 0.25 db/km. As the wavelength decreases still more, the water-vapor absorption decreases slightly and then rises sharply to another peak of nearly 18 db/km at 0.165 cm wavelength. An oxygen-absorption peak nearly as great occurs at a wavelength of 0.5 cm and another at 0.25 cm.

It should be pointed out that it is possible to take advantage of these high absorption bands as another method of providing a secure point-to-point communication link. If highly directional antennas are used with only enough power to provide an adequate signal-to-noise ratio at the receiver, the choice of a frequency right on one of the above absorption peaks will insure that no detectable signal will exist at ranges much beyond the terminals of the link.

Scattering losses due to fog, drizzle, and rainfall are not significant at frequencies below 10,000 megacycles, but rapidly become significant at higher frequencies. The rate of attenuation is also a function of the intensity of rainfall and the size and distribution of water droplets. Even moderate rainfall may cause an attenuation of 1 db/km or more for wavelengths of less than 1 cm.

ANTENNA APPLICATIONS

Since every electronic system that employs radiated electromagnetic energy to transmit or collect intelligence must employ one or more antennas, it is clear that every possible antenna type cannot be discussed here in detail. However, general classes of antennas that have wide application in electronic warfare will be defined and some general principles will be outlined.

TRACKING ANTENNAS

Whether used by gun-laying, interceptor, missile-, or target-tracking radar the requirements for tracking antennas are the same. The antenna must produce as narrow a searchlight or pencil beam as possible in order to provide a high order of resolution. Also, this concentration permits maximum range or greatest signal level at a given range. Usually, the beam must be nearly symmetrical and means must be provided to position or scan the beam to any desired direction. To provide as much anti-jam or anti-clutter capability as possible, side lobes must be kept to a minimum value. These requirements are met only by a symmetrical antenna having a uniform electrical phase distribution across the entire aperture. Mechanical gimbals provide the scanning required in most cases, although electronic techniques can be employed.

Conventional parabolic antennas for tracking applications meet most of the above requirements. The paraboloid has the property of producing a plane wave (uniform phase distribution) if a source of energy is located at the focus. Design considerations have been almost completely worked out for all parabolic antennas. Since any practical feed is not actually a point source but is of finite size and has its own radiation pattern, many variations of reflector feed combinations are employed. Attention must be given to the desired polarization, side-lobe level, available volume, required beam width, and the necessary collecting aperture. The feed may be excited by transmission lines passing through the reflector on the axis of the paraboloid or around the edge of the dish along a supporting arm.

There is, then, considerable judgment to be exercised in the detail design of conventional parabolic dishes, but the vast available backlog of experience in this type of antenna makes the task quite straightforward. It should be observed that parabolic dishes are used in small air-to-air homing missiles in either active or semi-active systems, in the interceptor that launches the missiles, in the ground radar guiding the interceptor to its target, and in many other instances. Thus, parabolic dishes from a few inches to hundreds of feet in diameter have application in electronic warfare, the size being a function of the wavelength used, the beam width desired, and the allowable space.

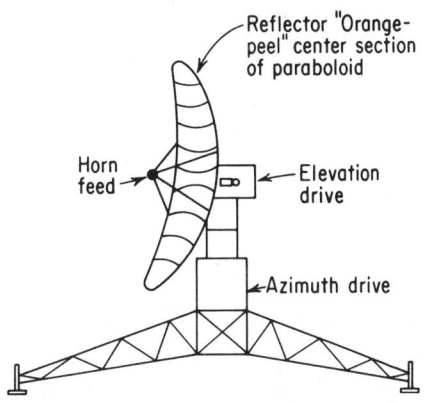

Figure 6-9. *Height-finding antenna with broad azimuth pattern and narrow elevation pattern.*

Variations of the symmetrical paraboloid for applications that do not require symmetry in the pattern usually take the form of parabolic sections. For example, in determining target height, broad azimuth and narrow elevation beams are required. Using the center section of a dish, as shown in Fig. 6-9, provides an antenna with such a pattern. In many cases the antenna rotates in azimuth as it nods in elevation.

Rotating electrical joints can be eliminated by using offset feeds and sections of paraboloids, as shown in Fig. 6-10. In antenna systems of this type the polarization is a function of relative reflector-feed orientation.

In addition to the target-acquisition function, error signals must be provided when the target is being tracked. A wide variety of techniques for generating appropriate signals have been developed, but the principles are

146   THE ROLE OF ANTENNAS

similar in all cases In essence, the fact that target illumination, and hence the returned signal, varies with the position of the target in the antenna pattern, is used to determine the target location in relation to the antenna axis. When the target echo moves off the antenna axis an error signal is generated and is used to correct the antenna bearing.

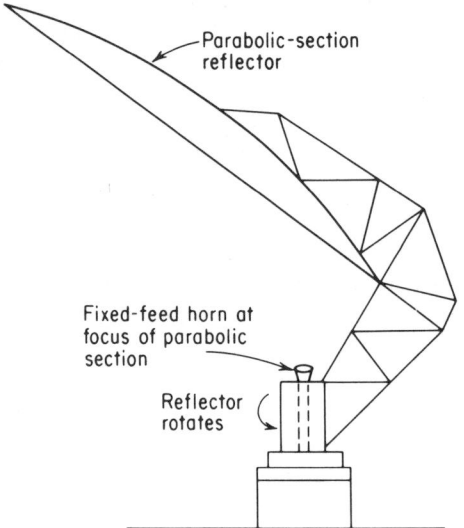

Figure 6-10. *Offset feed eliminates rotary joints.*

MANUAL TRACKING

The simplest form of this principle is the one used by ground-to-air searchlight operators, tracking on the beam maximum. A radar operator can manually position the antenna for maximum response as determined by an audible tone or by a signal on a scope. Although simple in principle and practice, this method is low in accuracy and slow in response to violent target maneuvers, because of the limitations of the human operator.

MULTIPLE-BEAM TRACKING

Next in order of complexity is the use of multiple beams. Here a number of offset overlapping beams provide target signals. By comparing signals received by all the beams and relating them to the known relative positions of the beams, it is possible to compute target direction to a degree of accuracy much greater than that obtainable from a single-beam system.

CONICAL SCAN

From the multiple-beam arrangement it is a logical step to consider the conical-scan tracking system employed in many fire-control and guidance

CONICAL SCAN 147

radars, already discussed in some detail in Chapter 5. The logic of this step derives from the fact that a conical-scan system provides an infinite number of beams by causing a single offset beam to rotate rapidly about the antenna axis. Figure 6–11 illustrates the principle of beam rotation as viewed in the far field of the antenna.

Beam rotation about the antenna axis may be accomplished by displacing the feed slightly from the axis and rotating it at the conical-scan frequency. Alternatively, the reflector can be tilted slightly and rotated in this off-axis

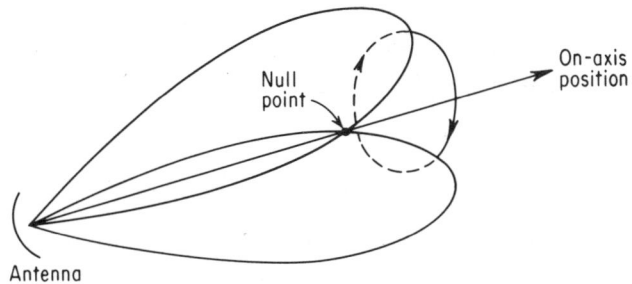

Figure 6-11. *Conical scan. Antenna beam describes a circle about the "on-axis" position.*

position. Both techniques are used, but the latter is useful only for small antennas.

The modulation produced on the target signal by an off-axis target is compared with the periodic voltage associated with the rotating feed or reflector. In an automatic tracking system this comparison produces error signals used to drive servos that position the antenna to eliminate the target-signal modulation. An absence of target-signal modulation indicates an on-axis target. When on-axis, the target is at the same point on each of the infinite number of beams and thus returns a constant signal to the radar.

MONOPULSE TRACKING

Since error-signal information rate from a conical-scan system is dependent on the mechanically limited conical-scan rate, other techniques have been developed to provide higher information rates. Also, since the conical-scan frequency can be detected and utilized in jamming by the target, a tracking technique was developed that cannot be detected by the enemy. The general name "monopulse" has been adopted to describe a family of such tracking systems. In these systems, both range and angle information are derived from each pulse returned from the target.

In a monopulse system the antenna must provide the equivalent of five beams positioned as indicated by Fig. 6–12. On each of the five beams the antenna and associated feed circuits must provide outputs from the pulses scattered by the target. Target-range information is determined from the center beam in the conventional manner. Elevation error signals are obtained by comparing the signals received by the two elevation beams, and azimuth

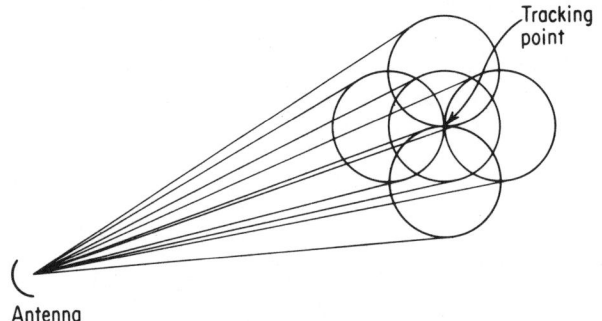

Figure 6-12. *Monopulse beam arrangement.*

error signals are determined in the same way. Transmission uses only the on-axis beam to provide maximum target illumination.

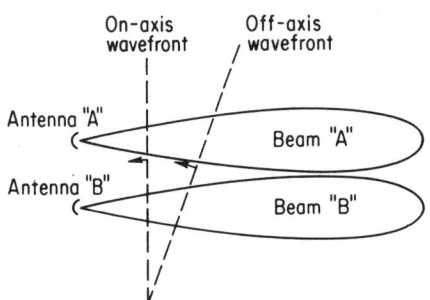

Figure 6-13. *Phase comparison monopulse.*

Comparison of the signals received from pairs of receiving beams, as mentioned above, is accomplished by either an amplitude or a phase-comparison technique. Although in actuality neither technique is purely phase or amplitude comparison, it is common practice to refer to either one or the other as the receiving principle. The phase-comparison system employs two displaced feeds. Error-signal generation depends on the fact that, although amplitude patterns associated with the two feeds are identical, the phase patterns differ. Thus, as seen in Fig. 6-13, signals arriving from off-axis directions will arrive in different phases at feeds A and B.

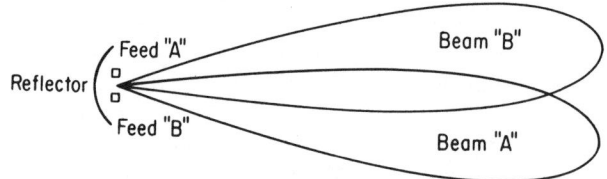

Figure 6-14. *Amplitude comparison monopulse.*

Amplitude-comparison systems employ two feeds located at the same point but producing patterns squinted off-axis in opposite directions. From Fig. 6-14 it can readily be seen that off-axis signals will produce outputs from

Feeds *A* and *B* that differ in amplitude by the difference in amplitudes of the two patterns in the direction from which the signal arrived. By properly designing the feeds, reflectors, and other microwave circuitry, it is possible to provide linear error signals in the region close to the axis. These error signals are then used to drive null-seeking servos that properly position the antenna and provide target-tracking control commands for the antenna or for homing missiles, if this is the ultimate intent.

SEARCH ANTENNAS

In the previous discussion of tracking antennas, it was tacitly assumed that the antenna was pointed at the signal to be tracked, whether it was a radar antenna receiving a reflected signal from a target, or was merely a passive listening antenna receiving radio signals from a moving vehicle or aircraft. The task of getting the tracking antenna "on target" in the first place is just as difficult as tracking the signal after acquisition. The use of narrow-beam, high-gain search antennas to provide maximum detection range is at variance with the desire to use wide-beam antennas to insure maximum probability of intercept and minimum time of search to cover the required volume.

EFFECT OF ANTENNA CHARACTERISTICS ON INTERCEPT PROBABILITY

Further consideration can now be given to the probability of intercept for the horizontal search mode as introduced in Chapter 4 (Fig. 4–2). In this treatment the target antenna will no longer be considered as a point source, and the scan rates of both the search and target antenna systems will now be included in the analysis. In order to illustrate the effect of these parameters on the problem, let the following situation be defined.

It will be assumed that a reconnaissance aircraft is searching for enemy early-warning radar sites while flying over friendly territory near the enemy border. The search antenna is a broad-band narrow-beam antenna, scanning in the horizontal plane. The maximum gain is directed at the horizon, since all the propagation factors considered previously (smooth-earth diffraction, tropospheric scatter, and ducting), result in maximum detection range for ground-based signals when the antenna beam is directed at grazing incidence to the Earth at the horizon.

It will be assumed that the radar antenna has a horizontal beam width of θ_1, and rotates with a period of t_1, whereas the search antenna has a beam width of θ_2 and a period of $t_2 < t_1$. The probability of intercept during one scan period of the radar antenna is given by:

$$P = \frac{\theta_1 t_1}{360 t_2}. \qquad (6\text{–}10)$$

The cumulative probability of intercept for two scans is:

$$P + (1 - P)P, \tag{6-11}$$

and for n scans, the cumulative probability is given by the series:

$$P_c = P + (1 - P)P + (1 - P)^2 P + \ldots (1 - P)^n P. \tag{6-12}$$

This is a series in the form of a geometric progression, and the sum may be expressed as

$$P_c = 1 - (1 - P)^n. \tag{6-13}$$

Substituting $\theta_1 t_1 / 360 t_2$ for P, and noting that n, the number of scans, is equal to the total time of looking T, divided by the period for each scan t_1, the cumulative probability of detection during the time T is:

$$P_c = 1 - \left(1 - \frac{\theta_1 t_1}{360 t_2}\right)^{T/t_1}. \tag{6-14}$$

Equation 6–13 can be written in the form

$$P_c = 1 - [(1 - P)^{-1/P}]^{-nP}. \tag{6-17}$$

Since P, the probability of intercept during one scan, is small, then P_c may be approximated by:

$$P_c = 1 - e^{-nP}. \tag{6-16}$$

Since $$(1 + P)^{1/P} = e \tag{6-17}$$

the cumulative probability is now approximated by:

$$P_c = 1 - e^{-\theta_1 T / 360 t_2}. \tag{6-18}$$

It will be noted that the cumulative intercept probability approaches unity as t_2, the period of the search antenna approaches zero. In other words, the faster the search antenna rotates the more certain the interception becomes. The search-antenna speed cannot be increased without limit, however, because of mechanical limitations, but, still more important, the search antenna must "look" at the radar beam long enough to receive at least one pulse from the radar. The time that the search antenna requires to scan its beam width θ_2 past any given point is $t_2 \theta_2 / 360$. In order to assure the reception of at least one pulse, this time must be at least equal to the pulse period of the radar. Thus:

$$\frac{t_2 \theta_2}{360} \geq \frac{1}{PRF}. \tag{6-19}$$

Consideration of Eq. 6–19 would lead one to the conclusion that to insure the greatest probability of intercept the search antenna should have as broad a beam as possible. In the limit this would be an omnidirectional antenna, i.e., $\theta_2 = 360$ degrees. This coverage is not the only consideration, however, since increasing θ_2 results in decreased antenna gain, and consequently in reduced range of intercept. Furthermore, the requirement that at least one pulse be received is certainly an optimistic assumption. In practice it might be necessary to receive ten or more pulses to insure a positive identification of the signal and to enable the pulse characteristics to be measured or recorded.

As an example, assume that the structural and aerodynamic characteristics of the airplane will permit the installation of a parabolic search antenna no greater than 3 ft in diameter, and that the flight mission is to search for radars operating at frequencies near 1000 megacycles and with *PRF*'s around 400 pulses per sec. From Eq. 6–6 it will be found that the beam width of the 3 ft parabola at 1000 megacycles is approximately 20 degrees. If ten pulses are required for a successful intercept it will be seen from Eq. 6–19 that

$$\frac{t_2 \times 20}{360} \geqq \frac{10}{400}. \qquad (6\text{--}20)$$

Thus $t_2 \geqq 0.45$ sec, and the scan rate for maximum probability of intercept should be 133 rpm.

DESIGN OF LARGE GROUND-BASED ANTENNAS

Although considerable emphasis has been placed upon the use of airborne antennas in electronic warfare, ground-based antennas are also of great value for both electronic reconnaissance and jamming pruposes. Very large, high-gain ground-based antennas can detect signals, via tropospheric scatter, from aircraft, missiles, and ground stations far beyond the horizon. Such signals can be monitored and recorded over long periods of time in order to obtain data on the enemy electronic order of battle, or if necessary, jamming transmitters can be used with high-gain antennas to blank out specific geographic areas.

Satellite tracking, deep-space communications, and radio astronomy, as well as ground-based antennas for electronic warfare, require very large antenna structures in order to obtain the maximum possible gain. The mechanical and structural problems associated with the design and fabrication of very large antennas are usually the limiting factors that affect the antenna performance.

A parabolic surface must be accurate to within $\pm \frac{1}{16}$ of a wavelength if it is to perform properly as an antenna. Consequently, the tolerance that the mechanical designer can guarantee for the surface of a large parabolic reflector will establish the maximum frequency that can be used for that particular antenna.

Let it be assumed that a 200 ft diameter reflector can be built with a surface tolerance of $\pm$ 0.01 ft, that a 600 ft reflector would have a tolerance of ± 0.033 ft, and that a 1000 ft reflector could be held to ± 0.1 ft. This variation of tolerance with reflector size is plotted in the lower curve of Fig. 6–15. The upper curve of Fig. 6–15 is obtained by simple calculation, based on the assumption that 16 times the surface tolerance is the minimum useable wavelength for that reflector.

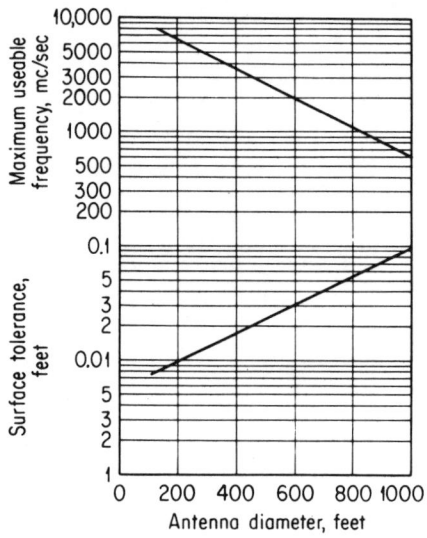

Figure 6-15. *Very large antenna design parameters.*

By substituting C/F for λ in Eq. 6–6 we obtain:

$$\theta = 70C/DF$$

where θ = half-power beam width,
C = velocity of light,
D = antenna diameter,
F = frequency.

Furthermore, the lower curve in Fig. 6–15 may be represented by the following expression:

$$\log F = -KD + A \quad (6\text{--}22)$$

where K = the slope of the curve
and A = the Y intercept.

Taking the log of both sides of Eq. 6–21 and substituting for $\log F$ from Eq. 6–22 we obtain:

$$\log \theta = \log(70C) - \log D + KD - A.$$

Differentiating with respect to D:

$$\frac{d \log \theta}{dD} = \frac{-0.434}{D} + K.$$

Setting the derivative equal to zero, we find that:

$$D = \frac{0.434}{K}. \quad (6\text{--}23)$$

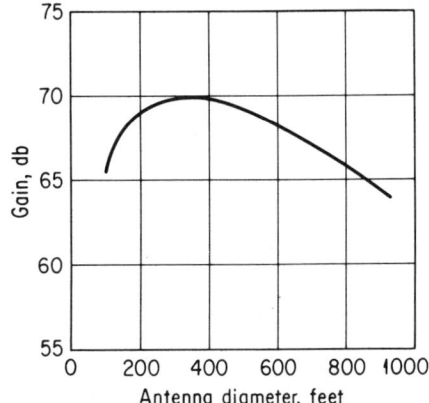

Figure 6-16. *Antenna gain at maximum useable frequency vs. antenna diameter.*

Thus it is found that a minimum beam width exists for an antenna diameter equal to $0.434/K$. K can be determined from Fig. 6–15 and is equal to 0.00125. Thus, $D = 0.434/0.00125 = 347$ ft.

This result indicates that if the original assumption concerning the accuracy with which large parabolic reflectors can be constructed is correct, then an antenna 347 ft in diameter will have the narrowest possible beam width

LARGE GROUND-BASED ANTENNAS 153

at its maximum useable frequency (400 megacycles/sec). It will also have the largest possible gain, since gain is inversely related to beam width. Figure 6–16 shows a plot of gain versus antenna diameter, and it is noted that the gain does pass through a maximum at 347 ft.

ANTENNAS OF OTHER TYPES

To the experienced antenna engineer it will be obvious that many sophisticated antenna types have not been discussed here. It is not our purpose in this chapter to provide either an encyclopedia of antennas or a textbook on antenna design, but merely to introduce the electronic-warfare specialist to some of the basic antenna characteristics with which he should be familiar.

7

OPTIMIZATION—CONSTRAINTS AND INCOMPLETE INFORMATION

OBJECTIVES OF AN ECM SYSTEM

Electronic countermeasures are of importance only in relation to the extent to which they expedite the accomplishment of a mission. They must function not as an entity in themselves but as an integral part of a system. Those concerned with the design of ECM must be fully aware of the relationship between the ECM equipment and the system of which it is a part in order to provide effective hardware. In this chapter we shall investigate certain aspects of the relationships and their application to ECM design and employment. In the first section we shall consider the methods of determining the optimum characteristics of ECM equipment subject to constraints imposed by the portion of the weight and volume capacity of the ECM carrier devoted to the equipment. The influence of the accuracy of knowledge of enemy equipment characteristics will also be examined. These considerations are of particular importance in the design and development phases of an ECM program. In the second part of the chapter we shall look into the proper allocation of available resources to ECM and weapons. This problem was briefly introduced in Chapter 1 and will now be considered in detail. This subject is of primary importance when procurement decisions must be made.

Whenever the discussion requires the consideration of a specific mission, it is assumed that the ECM is intended to enhance the effectiveness of an attack on targets embedded in a defense system. Accordingly, whenever reference is made to ECM equipment it is the friendly ECM that is under consideration. Whenever defense is considered it is with reference to the

OBJECTIVES OF AN ECM SYSTEM   155

enemy defense; the discussion of effectiveness also applies to the enemy defense. Thus, the effectiveness of the ECM is only implied and is actually measured by the degree to which it reduces defense effectiveness.

It is implicitly assumed that defense effectiveness in the presence of a given ECM configuration is known if the characteristics of the defense and the ECM are known.

OPTIMIZATION OF ECM EQUIPMENT CHARACTERISTICS

Attention will now be directed to determining the optimum characteristics (e.g., power, frequency range, and antenna directivity) of an ECM equipment or family of equipments. There are three considerations that must be observed in accomplishing the optimization: constraints, functional relationships between system characteristics and system effectiveness, and knowledge of the enemy's equipment. Constraints arise because the ECM equipment is only one component of the total attack system. It is restricted in size and weight by the characteristics of the vehicle in which it is borne. If there were no constraints on the size and weight of the ECM equipment, the optimum equipment would have infinite volume and infinite weight, because the larger the equipment, the more effective it can be made against a given electronic system or the greater the flexibility it will have in coping with the contingencies that arise in the mission. The constraints on the ECM equipment are not necessarily equal to the maximum capacity of the bearing vehicle, since the ECM may be only a part of the useable load. The problem being treated at this point may be stated as follows: given a certain fraction of the vehicle's payload devoted to ECM equipment, what should the equipment characteristics be in order to minimize enemy defense effectiveness?

The analytical procedure for optimizing, subject to constraints, requires the following definitions. Assume that there are m characteristics that uniquely define the ECM equipment. Let x_i be the ith characteristic (e.g., power), where i has the values $1, 2, ..., m$, and m represents the number of characteristics. The effectiveness of the ECM equipment is E, where E is a function of the x_i, that is:

$$E = E(x_1, x_2, ..., x_i, ..., x_m) \quad \text{effectiveness} \quad (7\text{-}1)$$

in addition, the weight, W, and volume, V, are functions of the x_i, thus:

$$W = W(x_1, x_2, ..., x_i, ..., x_m), \quad \text{weight} \quad (7\text{-}2)$$

and

$$V = V(x_1, x_2, ..., x_i, ..., x_m). \quad \text{volume} \quad (7\text{-}3)$$

The weight and volume are each constrained so that the weight cannot exceed some maximum value, w, and the volume cannot exceed some maximum value, v. Algebraically,

$$W(X) \leq w, \quad \text{weight constraint} \quad (7\text{-}4)$$

$$V(X) \leq v, \quad \text{volume constraint} \quad (7\text{-}5)$$

where X designates an m dimensional column matrix containing a dimension for each of the x_i; that is:

$$X = \begin{vmatrix} x_1 \\ x_2 \\ \cdot \\ \cdot \\ \cdot \\ x_m \end{vmatrix}. \quad (7\text{-}6)$$

The problem, then, is one of minimizing the defense effectiveness (Eq. 7-1) with respect to X, subject to the constraints (Eqs. 7-4 and 7-5), when the relationships (Eqs. 7-1, 7-2, and 7-3) are known. To visualize the problem, consider Fig. 7-1, where it is assumed that only a weight constraint exists.

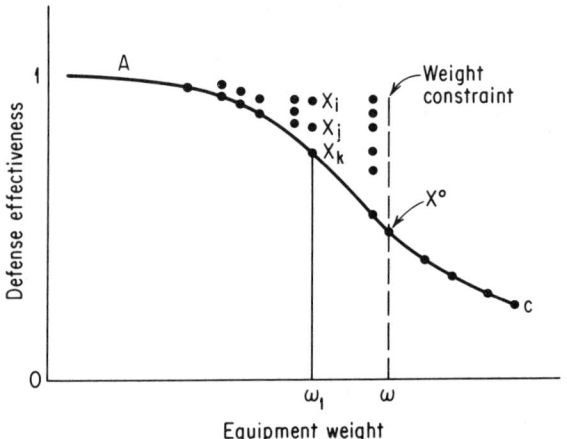

Figure 7-1. *Optimization of ECM equipment characteristics subject to a weight constraint.*

The figure shows a two-dimensional space of effectiveness and weight. For every set of equipment characteristics (X) there exists a point in the space. For any given value of weight (W) there is a continuum of values of E and obviously only the X yielding the minimum effectiveness for any given weight is of concern. For example, X_i, X_j, and X_k all have the weight W_1, but X_k results in the lowest defense effectiveness. Thus, X_i and X_j are inferior to X_k and need not be considered further. These minima yield a curve A–C which is a non-increasing function of weight. It is the intersection of this curve with the ordinate drawn through the weight constraint, w, that defines the maximum effectiveness and the set of optimum equipment characteristics:

$$X^0 = \begin{vmatrix} x_1^0 \\ x_2^0 \\ \cdot \\ \cdot \\ \cdot \\ x_m^0 \end{vmatrix}. \qquad (7\text{--}7)$$

Thus, in this single constraint situation, the condition on the weight (Eq. 7–4) becomes:

$$W(X) = w, \qquad (7\text{--}8)$$

that is, the inequality of condition (Eq. 7–4) is ignored and only the case in which the equipment weight is equal to the maximum tolerable value of weight, w, is of concern. Several methods exist by which the set of system characteristics X^0 that results in the minimum defense effectiveness and which has a weight w, can be determined. Among them is the method of the Lagrange undetermined multiplier.* In essence the method allows for the solution of problems having m unknowns (the set of m characteristics, which define the optimum set), and $m + 1$ conditions; m conditions arise from the requirement that the derivative of the effectiveness function with respect to each of the characteristics vanish, and one condition arising from the requirement that the weight must equal the weight constraint, w, by conventional calculus techniques. The method of solution is to define a function, $F(X)$, of X, which is obtained by adding $b[W(X) - w]$ to the defense system effectiveness, i.e.,

$$F(X) = E(X) - b[W(X) - w] \qquad (7\text{--}9)$$

where b is the undetermined multiplier. It is apparent from Eq. 7–9 that $F(X)$ is equal to $E(X)$ whenever Eq. 7–8 is satisfied; thus if $F(X)$ is minimized, $E(X)$ is also minimized for any value of X that satisfied Eq. 7–8. By employing $F(X)$ rather than $E(X)$ as the function to be minimized, the multiplier, b, provides the needed $(m + 1)$th unknown. Formally, the procedure consists of solving the $(m + 1)$ equations:

$$\frac{\partial F}{\partial x_i} = 0$$

$$W(x) = w,$$

subject to the usual conditions for a minimum:

$$\frac{\partial^2 F}{\partial x_i \partial x_j} > 0.$$

* See Bliss, G. A., *Lectures on the Calculus of Variations*, The University of Chicago Press, Chicago, 1946.

The technique will in general yield several local minima, and these must be compared to determine the absolute minimum. As an illustration, consider the following example.

Let the effectiveness of the defense be given as:

$$E(X) = \frac{100}{x_1 x_2 x_3}$$

and let the weight be given by

$$W(X) = 2x_1 + x_2 + 4x_3$$

and let the weight constraint be

$$w = 15.$$

Then:

$$F(X) = \frac{100}{x_1 x_2 x_3} - b(2x_1 + x_2 + 4x_3 - 15)$$

$$\frac{\partial F}{\partial x_1} = \left[\frac{-100}{x_1 x_2 x_3} \cdot \frac{1}{x_1} - 2b\right] = 0$$

$$\frac{\partial F}{\partial x_2} = \left[\frac{-100}{x_1 x_2 x_3} \cdot \frac{1}{x_2} - b\right] = 0$$

$$\frac{\partial F}{\partial x_3} = \left[\frac{-100}{x_1 x_2 x_3} \cdot \frac{1}{x_3} - 4b\right] = 0.$$

Solution of the above yields

$$2x_1 = x_2$$
$$4x_3 = x_2.$$

Substituting into the weight equation, $W(X)$, and setting this equal to the constraint, 15, yields optimum values of the x_i as follows:

$$x_1 = 2.5$$
$$x_2 = 5$$
$$x_3 = 1.25$$

and a minimum defense effectiveness of:

$$E(X) = 6.4.$$

The simplified problem of a single constraint provides an entry into the original problem in which two constraints, weight and volume, exist. Refer

OPTIMIZATION OF ECM CHARACTERISTICS   159

to Fig. 7-2, which indicates a three-dimensional space of defense effectiveness, weight, and volume. For any value of X there exists a point in this space. In general, for any given value of weight and volume, there is a continuum of values of effectiveness. That is, there are a multitude of values of X corresponding to a particular weight and volume, and each value has, in general, a different value of effectiveness associated with it. Here, as in the case of the single constraint, the only concern is with the value of X for which

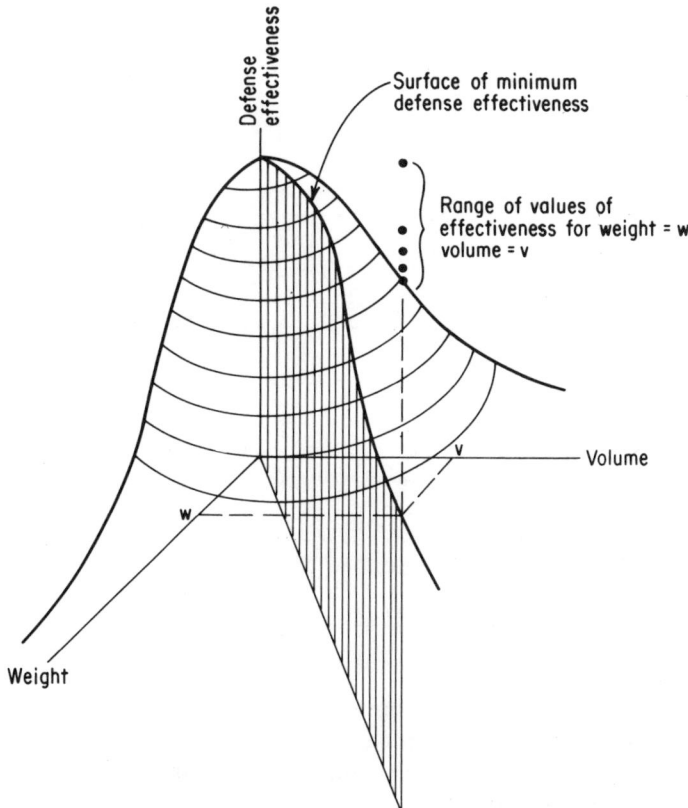

Figure 7-2. *Effectiveness as a function of weight and volume.*

the value of effectiveness is a minimum for any pair of values of weight and volume. Thus there may be conceived a surface in the weight-volume-effectiveness space. Each point on the surface has a minimum value of defense effectiveness for the indicated pair of weight and volume values and each corresponds to some value of X. To arrive at a solution it is first necessary to realize that the optimum equipment will have either a maximum allowable weight or a maximum allowable volume. This limit can be seen by choosing some point in Fig. 7-3 for which both the volume and weight are less than the constraining values. The weight can be increased and the defense effectiveness decreased by varying the value of an appropriate variable, x_i. As the weight is increased, the volume will in general increase so that the weight and volume point will describe a path similar to either a–b or a–c. The weight may increase until either the weight or the volume becomes equal to its respective constraining value, w or v. At this point the defense effectiveness

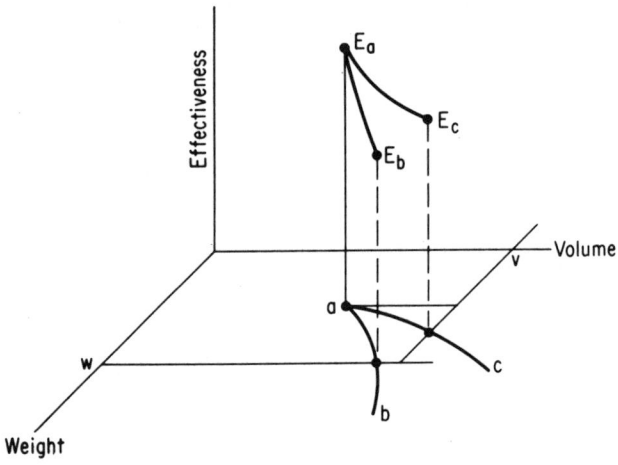

Figure 7-3. *For optimum equipment either weight or volume equals restraint.*

will in general be less than it was when the values of weight and volume were at a point within the rectangle described by the lines:

$$W = 0 \quad V = 0$$
$$W = w \quad V = v$$

Having established that at optimum either the weight or the volume equals the respective constraint, the procedure for optimization under conditions in which the Lagrange multiplier is applicable becomes quite straightforward. The procedure is outlined below.

A. Assume that the weight constraint predominates, i.e.:

$$W(X) = w.$$

1. By the procedure given under the single constraint situation, determine the values of the equipment characteristics where the defense effectiveness has local minima.
2. Determine the defense effectiveness for these values, E_w^j where j corresponds to the jth local minimum subject to the weight constraint.
3. Compute the resulting volumes, which will be termed V_w^j.

B. Assume that the volume constraint predominates, i.e.:
$$V(X) = v.$$
1. By the procedure given under the single constraint situation, determine the values of the equipment characteristics where the defense effectiveness has local minima.
2. Determine the defense effectiveness for these values, E_v^k, where k corresponds to the kth local minimum subject to the volume constraint.

C. Assume next that both the weight and volume constraint apply, i.e.:
$$V(X) = v$$
$$W(X) = w.$$
The solution is accomplished in this case by defining the function to be minimized as:
$$G(X) = E(X) - b[V(X) - v] - c[W(X) - w] \qquad (7\text{-}10)$$
where b and c are both Lagrange undetermined multipliers. Formally, the procedure is accomplished by satisfying the following $m + 2$ conditions:
$$\frac{\partial G(X)}{\partial x_i} = 0 \qquad V(X) = v \qquad W(X) = w \qquad (7\text{-}11)$$
together with the conditions that the second derivatives must all exceed zero. Equations 7–11 yield a set of local minima $E_{w,v}^l$.

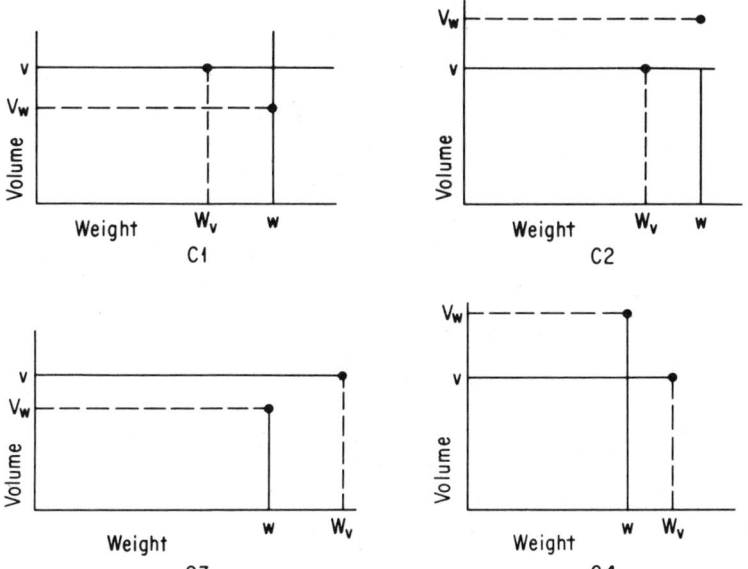

Figure 7-4. *Optimization conditions for weight and volume constraints.*

D. In this step the local minima will be examined to determine the absolute minimum subject to the constraints (Fig. 7-4).

1. Compare the V_w^j (Step A3) with v. Disregard those j's for which
$$V_w^j > v.$$

2. Compare the W_v^k (Step B3) with w. Disregard those k's for which
$$W_v^k > w.$$

3. Compare the values of the E_w^j for all j's remaining after Step D1, the values of the E_v^k for all k's remaining after Step D2 and the values of E_{wv}^l (Step C). Disregard all of each set except that for which the effectiveness (E_w^j, E_v^k, E_{wv}^l) is a minimum. Let these be designated as $E_w^{j^0}$, $E_v^{k^0}$, and $E_{wv}^{l^0}$ respectively; that is

$$E_w^{j^0} = \min_j E_w^j$$
$$V_w^j \leq v$$
$$E_v^{k^0} = \min_k E_v^k$$
$$W_v^k \leq w$$
$$E_{wv}^{l^0} = \min_l E_{wv}^l.$$

4. Of the three values ($E_w^{j^0}$, $E_v^{k^0}$, $E_{wv}^{l^0}$), select that which is the smallest. The corresponding local minimum (j^0, k^0 or l^0) is the absolute minimum subject to the constraints. The values of the x_i at this local minimum are the optimum system characteristics.

Our discussion of optimization techniques subject to weight and volume constraints is completed. The following example is intended to provide clarification.

Let the effectiveness of the defense be:

$$E(X) = \frac{100}{x_1 x_2 x_3}$$

the weight be:

$$W(X) = 2x_1 + x_2 + 4x_3$$

and the volume be:

$$V(X) = 5x_1 + 10x_2 + 20x_3.$$

Furthermore, let the constraints be:

$$\text{weight constraint } w = 15$$
$$\text{volume constraint } v = 50.$$

(a) As given in the example for the single constraint:

1. $x_1 = 2.5$
    $x_2 = 5$
    $x_3 = 1.25$
2. $E(w) = 6.4$
3. $V(w) = 87.5$.

(b) Assuming that the volume constraint applies,

1. $x_1 = 3\tfrac{1}{3}$ $x_2 = 1\tfrac{2}{3}$ $x_3 = \tfrac{5}{6}$
2. $E_v = 21.6$
3. $W_v = 11\tfrac{2}{3}$.

(c) Then, since $V_w > v$ and $W_v < w$ the procedure of Part C3 indicates the optimum values are those obtained when the volume constraint applies, i.e., that:

$$x_1 = 3\tfrac{1}{3} \quad x_2 = 1\tfrac{2}{3} \quad x_3 = \tfrac{5}{6}$$

and

$$E^0 = E_v = 21.6.$$

Once again, it is well to realize that the Lagrange multiplier may be applied only under certain conditions. A more general approach to the solution of this type of problem is that of dynamic programming.*

DEFINITION OF THE EFFECTIVENESS FUNCTION

Optimization of ECM equipment characteristics requires knowledge of the relationship between defense-system effectiveness and ECM equipment characteristics. In general, if a given defense system is postulated, the required relationships can be defined, although even this is no trivial problem. For example, consider a local defense employing surface-to-air missiles. If these missiles require accurate range information in order to be effective, the equations presented in Chapter 5 (Eqs. 5–64 to 5–66) for the burn-through range indicate the distance from the defense radar that a fire-control solution can be commenced. This information requirement, together with defense time delays, distance vs. time curves of the missile, and aircraft speed determine the number of salvoes that the defense may fire. However, the problem is complicated because in some cases range may be accomplished by triangulation or the defense may be capable of an alternate home-on-jam mode; thus, results based on burn-through alone may grossly underestimate the defense.

* Bellman, Richard, *Dynamic Programming*, Princeton University Press, Princeton, N. J., 1957.

In addition, jamming may not deny raid-location information completely but may serve only to degrade the accuracy. All of these factors must be treated to arrive at a valid analysis. Once the analysis is accomplished it will only apply to the specific defense system defined by the characteristics assumed in the analysis. In general, however, these characteristics will not be known precisely. Errors in observation of the defense system, together with the variations in the defense-system characteristics in the time span from observation to employment of the ECM equipment lead to uncertainty. The errors in observation arise from inaccurate or inadequate electronic reconnaissance procedures (Chapter 4) and misleading and fragmentary intelligence data. The variations with time result from normal defense evolution as well as the enemy's intentional attempts to react to ECM developments. In defining the effectiveness function it is essential to admit the uncertainty with which the characteristics of the defense are known, and to adopt a function that will provide a hedge against the various possibilities. The problem may be visualized by referring to Fig. 7-5, which illustrates the effectiveness of two defense systems in the presence of three ECM system configurations. If the enemy is known to have System A, then ECM Type 1 would be the logical choice. Similarly, if he were known to have System B, then ECM Type 2 would be the best choice. However, if the uncertainty in our observation leads to the belief that he may have either System A or System B, the tendency would be to choose ECM Type 3 as a hedge against the uncertainty. Type 3 is not as effective as Type 1 against A or Type 2 against B, but in the presence of the uncertainty it would tend to be favored. It also appears reasonable to tend to select ECM Type 1 as the likelihood that System A will exist increases. It is this type of consideration that will be investigated in subsequent discussion. Our intent is to introduce the uncertainty in a manner sufficiently formal as to allow analytical treatment.

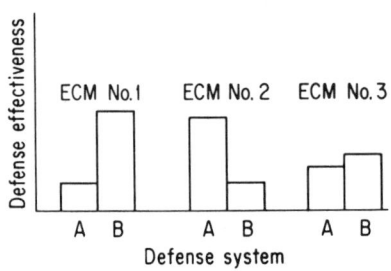

Figure 7-5. *Illustration of hedging against uncertainty.*

In practice, the situations that determine the treatment of uncertainty range from: (1) perfect knowledge is available because of very accurate reconnaissance and short ECM development time; to (2) only an approximate estimate of the electronic system configuration is available because of reconnaissance inaccuracy and moderately long development times; to (3) knowledge only that the enemy can choose any system configuration between some very broad limits, because of very poor reconnaissance or a long development time. Each of these situations will be examined.

The first situation, that of perfect information, will be treated as a special case of imperfect information.

THE EFFECTIVENESS FUNCTION

Consider a situation in which the enemy electronic system does not vary because of reaction with the ECM equipment. However, its characteristics are not known with certainty because of time delays and errors in reconnaissance. The treatment requires the notion of probability presented earlier. Let it be assumed that N assessments of the defense system had been made and that of these, N_j assessments indicated the existence of System j. Furthermore, assume that there is some general correlation between the observation of System j and the probability that System k exists. (E.g., if one of the factors is the number of transmitters, then if 100 transmitters are observed at some time and 50 at some other time, the probability that there are 50 transmitters is zero. The observation of 50 transmitters may have resulted from a lull in electronic activity.) It is essential to the balance of the discussion that we employ the results of the reconnaissance flights in order to estimate the probability that a particular electronic system exists. Assume that through prior knowledge, whether it is our own state of electronic systems, or fragmentary intelligence, or a knowledge of the enemy state-of-the-art, it is estimated that the probability that the enemy system is of Type k is P^k. Further, because of the characteristics of System k and our reconnaissance program, it is estimated that if System k in fact exists, the probability that the characteristics of System j will be observed is p_{kj}. Our interest is in determining the probability that k exists if in N overflights the set $\{N_j\}$ is observed. Let this probability be:

$$Q(\{N_j\}, k). \tag{7-12}$$

It can be shown that:*

$$Q[\{N_j\}, k] = \frac{P^k P(k, \{N_j\})}{\sum_k P^k P(k, \{N_j\})}, \tag{7-13}$$

where $P(k, \{N_j\})$ is the conditional probability that if k exists the set $\{N_j\}$ will be observed and it in turn is given by the multinomial distribution:

$$P(k_j\{N_j\}) = N! \prod_j \left[\frac{p_{kj}^{N_j}}{N_j!}\right] \tag{7-14}$$

which can be substituted into Eq. 6–13 to yield:

$$Q[\{N_j\}, K] = \frac{P^k \prod_j [p_{kj}^{N_j}]}{\sum_k P^k \prod_j [p_{kj}^{N_j}]} \tag{7-13(A)}$$

Thus, by means of Eq. 7–13(A), the probability that each of the enemy electronic systems exists may be computed on the basis of the most recent reconnaissance data. It is to be noted that the terms $Q(\{N_j\}, k)$ replace the values of P^k in Eq. 7–13(A) for use with each newer set of reconnaissance observations. The procedure will be illustrated by a simple example.

* Feller, William, *An Introduction to Probability Theory and Its Application*, John Wiley and Sons, Inc., New York, 1950.

Example: Consider that the enemy may have one of two systems, 1 and 2, and that to the best of the a priori knowledge available they are equally likely. Thus,

$$P^1 = 0.5$$
$$P^2 = 0.5.$$

Further, let the probability that if 1 exists then 2 will be observed, be 0.10, and the probability that 1 will be observed be 0.90, and similarly if 2 exists. That is,

$$p_1(2) = p_2(1) = 0.10$$
$$p_1(1) = p_2(2) = 0.90,$$

and that in four overflights 1 was observed three times and 2 was observed once. That is,

$$N^1 = 3$$
$$N^2 = 1$$
$$N = 4.$$

Then the probability that, if 1 exists, this set will be observed, is:

$$P(1, 3, 1) = 4 \cdot (0.9)^3 (0.1)^1 = 0.293,$$

and if 2 exists it is:

$$P(2, 3, 1) = 4.0(0.10)^3 (0.9)^1 = 0.0036,$$

which yields the new probability that 1 and 2 exist,

$$Q(2, 1, 1) = \frac{0.5(0.293)}{[0.5(0.293) + 0.5(0.0036)]} = 0.99 = P^1$$

$$Q(2, 1, 2) = \frac{0.5(0.0036)}{[0.5(0.293) + 0.5(0.0036)]} = 0.01 = P^2.$$

Thus, by virtue of the observations, the estimate that 1 and 2 were equally likely led to the estimate that System 1 is 99 times as likely to be the existing system than is System 2.

Having established the probabilities of the various enemy electronic systems, it is now possible to define the effectiveness function for this case of imperfect information. There are several ways to approach the problem, depending on whether the concern is with the expected effectiveness of the enemy system or with the probability that the effectiveness does not exceed some maximum tolerable level. If it is the expected effectiveness that is of interest, then the effectiveness function for an ECM system defined by the column matrix, X, against the enemy system is given by:

$$E(X) = \sum_j P^j E^j(X) \qquad (7\text{--}15)$$

where P^j is the probability that enemy System j will exist at the time of use of the ECM and $E^j(X)$ is the effectiveness of the ECM against System j.

The probability that the enemy system effectiveness will not exceed some maximum tolerable value, E_m, is computed by arranging the systems in order of increasing effectiveness in the presence of a given ECM equipment, as designated by X. The probabilities of all systems having effectivenesses not exceeding the maximum tolerable level are then summed, and this sum is the probability that enemy system effectiveness will not exceed the maximum level. This process must be repeated for each value of X that allows satisfaction of the weight and volume constraints. The value of X that minimizes the probability that enemy system effectiveness does not exceed E_m is optimum. Although in special cases the process may be accomplished with analytical techniques, it is in general one of cumbersome trial and error.

It is to be noted that if there is certainty that some defense system, J, will exist then the probabilities become

$$P^j = 0 \quad j \neq J$$
$$P^j = 1 \quad j = J.$$

Thus, if we are working with expected values, the effectiveness function becomes:

$$E(X) = E^J(X). \tag{7-16}$$

This same result occurs, of course, when we are considering the probability that the enemy defense does not exceed some given value. In this case, J is the only entry in the table.

Finally, we must consider the effectiveness function for a situation in which the enemy has the option of selecting his defense system from a broad range of alternatives. The only reliable information available to the ECM designer consists of the characteristics of the systems within the range. It is assumed that the enemy also has knowledge of our ECM state-of-the-art but not of our choice of system.

To enter the discussion, consider that either Defense System A or System B may be selected, and that ECM Equipment 1 or 2 may be designed. Let the effectiveness of Defense System A in the presence of ECM 1 be $E(A, 1)$ and, similarly, for System A in the presence of ECM 2 be $E(A, 2)$, B in the presence of 1 be $E(B, 1)$, and B in the presence of 2 be $E(B, 2)$. These symbols may be arranged in the form of an effectiveness matrix (Table 7-1).

Table 7-1. *Defense effectiveness in the presence of ECM*

Defense System	A	B
ECM Equipment		
1	$E(A, 1)$	$E(B, 1)$
2	$E(A, 2)$	$E(B, 2)$

168 OPTIMIZATION

In Table 7–2, values have been substituted for the symbols and they will be used as an illustration.

Table 7–2. *Illustrative effectiveness matrix*

Defense System	A	B
ECM Equipment		
1	3	6
2	10	8

It is assumed that the matrix is known to the defense as well as to the ECM designer. The defense by choosing System A would have an effectiveness of 10 if ECM 2 were employed; however, the effectiveness would be only 3 if ECM 1 were employed. Thus, he could choose equipment A and gamble that ECM 1 would be developed rather than 2. On the other hand, if System B were selected the defense could not attain the high level that could result from System A, but neither would the risk of the low level of 3 be a possibility. For System B, effectiveness would be 6 and 8 for ECM 1 and 2 respectively. The question then is whether or not the defense should gamble on the high level of 10 and risk a low level of 3 by selecting System A, or should he "play it safe" (if such a term can be used in this context), and choose System B? The answer, it is fairly obvious, is System B, if the ECM designer can be credited with reasonable logic. By selecting ECM Type 1, the defense effectiveness is less than that for Type 2 regardless of the choice of Defense System. Thus, the correct ECM selection is 1 and the correct Defense System is B. If Defense A is chosen, defense effectiveness would degrade from 6 to 3. If ECM 2 were chosen, defense effectiveness would increase to 8.

These considerations may be extended to a general case in which there are n possible defense systems and m possible ECM equipments.* This combination gives rise to an m by n matrix as indicated in Table 7–3, for which m is 4 and n is 5.

Table 7–3. *Four-by-five effectiveness matrix*

Defense System	A	B	C	D	E
ECM Equipment					
1	0	7	10	4	3
2	7	5	4	2	0
3	1	4	3	2	2
4	8	6	5	9	7

For this illustration the "best" choice of defense system and ECM equipment is not immediately obvious. In fact, the definition of what a "best" choice is relies on an intuitive concept. Consider the selection of any defense system, say System C for example. The enemy, by selecting C, may obtain an

---

* Although systems and equipments are discussed here, the considerations apply more generally to mixtures of various systems and various equipments.

effectiveness of 10 if ECM 1 is chosen. However, defense effectiveness is only 3 if ECM 3 is chosen. Thus, by choosing System C, an effectiveness of at least 3 and possibly as much as 10 may be obtained. On the other hand, if System E is chosen the effectiveness may range from 0 to 7. The defense is assured only of obtaining the minimum effectiveness in each column. On the other hand, the ECM designer by observing the maximum in a row can only be assured that the defense effectiveness will not exceed this amount if the corresponding ECM equipment is chosen. The defense systems and their assured minima are as follows: A, 0, B, 4, C, 3, D, 2, E, 0. The ECM equipments and the assured maxima are: 1, 10, 2, 7, 3, 4, 4, 9. From these we may see that the defense is assured of an effectiveness of at least 4 if B is chosen, and 4 is the maximum of the minimum assured values. Furthermore, the ECM designer can be assured of a defense effectiveness no greater than 4 if he chooses Equipment 3, and this is the minimum of all the maxima. The best choice for the defense is then said to be System B and the best choice of ECM equipment is 3. In more general terms, the best choice of defense system is that which maximizes the minimum possible effectiveness, i.e., which yields the maxi-min. The best choice of ECM equipment is that which yields the minimum of the maxima, i.e., the mini-max. For Table 7-3 it is to be noted that the mini-max and maxi-min are both 4. In this case a saddle point is said to exist and 4 is termed the value of the game. In general, a saddle point does not exist if only one type of ECM equipment or one type of defense system is considered for each entry. However, by forming combinations of equipments and combinations of systems, saddle points will always exist. It is to be noted in the example that if one party makes the "best" selection and the other does not, the one making the "best" selection derives an advantage in excess of the value of the game. Thus, if Defense System B is chosen and ECM Equipment 1, instead of 3, is selected, then the defense effectiveness is 7 rather than 4.

We may formalize the above discussion as follows:

Let $E(i,j)$ be the effectiveness of the jth defense system in the presence of the ith ECM equipment. Then the optimum defense system, j^0, and ECM equipment, i^0, are defined by:

$$E(i^0, j) \leq E(i^0, j^0) \leq E(i, j^0).$$

That is, $E(i^0, j^0)$ is at once the maximum in the row and the minimum in the column.

In the case just discussed, that in which only the range of possibilities of defense systems are known, the effectiveness function that is to be employed in the optimization of the ECM equipment is the effectiveness of the defense system having the greatest effectiveness of those in the range of possible defense systems, i.e.:

$$E(X) = \max_{j} E^j(X). \tag{7-17}$$

OPTIMUM DISTRIBUTION OF AVAILABLE PAYLOAD BETWEEN WEAPONS AND ECM

We now direct our attention away from the development to the employment of ECM. Consider the case of a cell of bombers that has been assigned to attack a given target in the event of war. Assume that the number of bombers in the cell is fixed at some value, N. Each bomber is capable of carrying a total weight and a total volume not exceeding some fixed values, and this payload may be distributed in any proportion between ECM and weapons. In attacking the target the cell must fly through a portion of the enemy's area and local defense. The problem is to define the allocation of aircraft payload between ECM and weapons in a manner that assures maximum probability that at least some given level of target destruction will be accomplished. Two cases may be considered. In the first the payloads of all bombers are the same; that is, the force is homogeneous. In the second, or heterogeneous case, each bomber will carry either bombs or ECM and the problem is to determine the number of bombers with each type of load. Only the homogeneous case will be treated. Since the discussion critically depends on the nature of the defense and the target-damage function, these factors will be discussed as preliminaries.

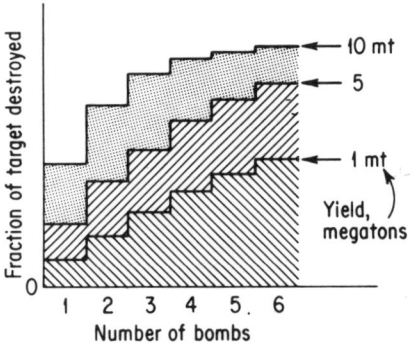

Figure 7-6. *Hypothetical damage function for a complex target.*

For a given target the fraction of destruction depends on the number of bombs dropped, the yield of the bomb, and the delivery accuracy. It will be assumed that the latter is known and is not a variable in the problem. The damage function has been developed for many complex targets in vulnerability studies at Rand, Stanford Research Institute, and Convair. A hypothetical example is presented in Fig. 7-6. There, the fraction of the target destroyed is presented as a function of the number of bombs dropped for various bomb yields. The function is, of course, discontinuous because of the requirement that the number of bombs be an integer. In the event that the target is small compared with the lethal radius and CEP, it may be treated as a point, which allows for the probability that the target will be destroyed if n bombs are delivered, to be computed simply by the following equation:

$$P_{(n)} = 1 - 0.5^{(R_L/CEP)^2} \cdot n \qquad (7\text{-}18)$$

where $P_{(n)}$ = the probability that if n bombs are dropped the target will be destroyed,

R_L = radius of destruction,

CEP = circular error probable (the radius of a circle drawn about the point of aim, within which 50% of the bombs will fall).

Furthermore, the lethal radius is given by an equation of the following general type:

$$R_L = K\frac{Y^{m_1}}{PSI^{m_2}}, \qquad (7\text{-}19)$$

where Y is the yield, PSI is the overpressure required to cause the desired level of damage and K, m_1, and m_2 are constants for a given target type over a

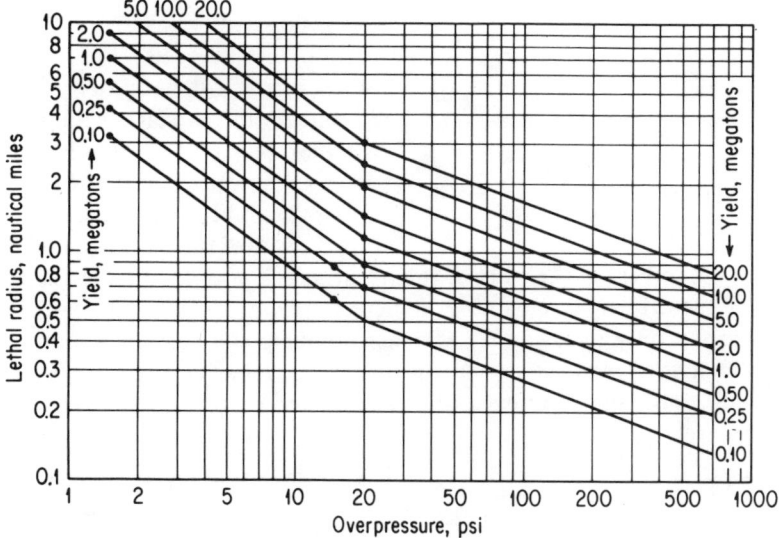

Figure 7-7. *Lethal radius as a function of overpressure and yield.*

range of Y and PSI. The curves of Fig. 7-7 allow direct reading of R_L and are more convenient than the formula.

The yield is directly related to bomb weight and the relationship is available in classified reports. With this relationship, the damage function (Fig. 7-6), and a decision as to the required damage level, the number of bombs of a given weight that must be delivered may be plotted. Figure 7-8 presents an illustrative example.

Now, consider the defense and its effectiveness in the face of ECM. The number of times that the cell will be attacked, M, must be determined by means of a penetration analysis. Next, the possible mixes of ECM and bomb weight that may be carried aboard an aircraft are determined. For example, the mix may consist of no ECM equipment and a 5,000 lb bomb; 1 ECM equipment and a 3,500 lb bomb; or 3 ECM equipments and a 500 lb bomb. Let k be the index of the mix, where k may signify the number of ECM equipments. Corresponding to each value of k there is a number of bombs that must be delivered, n_k, and n_k may be read from Fig. 7–8. The probability that a given attack by the defense will destroy a bomber is then a function of k and the number of bombers remaining at the time of the attack, j. Let $p(k, j)$ be the probability that an attack succeeds if mix k is employed and j bombers remain. Thus, for a cell of size N and for m possible mixes of ECM and bombs, $N \cdot m$ values of $p(k, j)$ must be computed. Once this computation is completed, the probability that the required damage will be obtained can be computed as follows. Assume that at the start of the ith attack by the defense, there are j bombers remaining in the cell. After the ith attack there will be either $j - 1$ or j bombers in the cell, depending on whether or not the attack by the defense succeeded. For the kth mix, the probability of there being j bombers after the ith attack if there were $j + 1$ bombers at the start of the ith attack is $p(k, j + 1)$ and the probability of there being j remaining if there were j at the start is $1 - p(k, j)$. These are the only two ways in which j bombers can remain after the ith attack. Therefore, we may write the following recurrence formula for the probability that j bombers exist after the ith attack for ECM bomb mix k, $P_i(k, j)$:

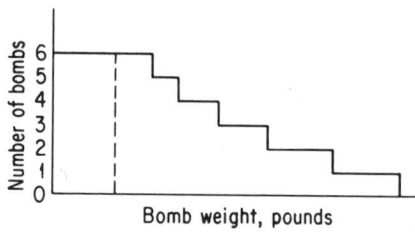

Figure 7-8. *Number of bombs of given weight required to accomplish a fraction, F, damage to target.*

$$P_i(k, j) = p(k, j + 1)P_{i-1}(k, j + 1) + [1 - p(k, j)]P_{i-1}(k, j). \tag{7-20}$$

In addition, the following boundary conditions must be observed:

$$P_i(k, j) = 0 \quad j < 0$$
$$P_i(k, j) = 0 \quad j < N - i \tag{7-21}$$
$$\sum_{j=N-i}^{N} P_i(k, j) = 1.$$

To illustrate the use of Eq. 7–20 let $N = 3$, $M = 2$, and the values of $p(k, j)$ be:

j	$p(k, j)$
3	0.2
2	0.4
1	0.6

Then,

$$P_1(k, 3) = [1 - p(k, 3)]P_0(k, 3)$$
$$= (1 - 0.2) \cdot 1 = 0.8 \qquad (1 - 0.2) \cdot 1$$

$$P_1(k, 2) = p(k, 3)P_0(k, 3)$$
$$= 0.2 \cdot 1 = 0.2$$

$$P_2(k, 3) = [1 - p(k, 3)]P_1(k, 3)$$
$$= (1 - 0.2) \cdot 0.8 = 0.64$$

$$P_2(k, 2) = p(k, 3)P_1(k, 3) + [1 - p(k, 2)] \cdot P_1(k, 2)$$
$$= 0.2 \cdot 0.8 + (1 - 0.4)0.2 = 0.28$$

$$P_2(k, 1) = p(k, 2)P_1(k, 2) + [1 - p(k, 1)]P_1(k, 1)$$
$$= 0.4 \cdot 0.2 + (1 - 0.6) \cdot 0 = 0.08.$$

The probability that at least the required number of bombs (n_k) arrive at the target if the cell is attacked M times is:

$$Q(k) = \sum_{j=n_k}^{N} P_M(k, j). \qquad (7\text{--}22)$$

Equations 7–20, 7–21, and 7–22 define the probability of a successful mission. The optimum mix, k^0, is that for which $Q(k)$, the probability of the required destruction, is a maximum. Formally, k^0 is defined by:

$$Q(k^0) = \max_{k} Q(k), \qquad (7\text{--}23)$$

where $k = 0, 1, ..., m$.

The procedure is vastly simplified if the probability of a defense success depends only on the ECM-bomb mix, k. In this case the $p(k, j)$ are replaced by p_k, denoting independence of the number of bombers remaining.

If the probability of success of a defense attack is independent of the number of attacks, the probability of j surviving after i attacks is given by the binomial distribution for $N - j$ successes in i trials, i.e.,

$$P_i(k, j) = \binom{i}{N-j} p_k^{n-j} (1 - p_k)^{i-n+j}, \qquad (7\text{--}24)$$

where

$$\binom{i}{N-j} = \frac{i!}{(i - N + j)!(N - j)!}, \qquad (7\text{--}25)$$

and the probability of a successful mission is the probability that $N - n_k$ or fewer of the M defense attacks succeeded, i.e.:

$$Q(k) = \sum_{0}^{N-n_k} \binom{M}{j} p_k^j (1 - p_k)^{M-j}. \qquad (7\text{--}26)$$

For convenience in computation, $Q(k)$ may be rewritten as:

$$Q(k) = 1 - E(N, N - n_k + 1, p_k), \qquad (7\text{--}27)$$

where

$$E(n, r, p) = \sum_{x=r}^{n} \binom{n}{x} p^x (1-p)^{n-x}$$

is the cumulative binomial distribution for which tables exist.

To clarify the procedure, consider the following example:

> The probability of a defense success is independent of the number of bombers in the raid; i.e., the defense is not saturated.
> The cell consists of 3 bombers, $N = 3$.
> Four attacks will be made, $M = 4$.
> The mixes, numbers of bombs required, and the defense kill probabilities are:

k	n_k	p_k
0	1	0.5
1	2	0.2
2	3	0.1

$$Q(0) = \sum_{0}^{2} \binom{4}{j} 0.5^j (0.5)^{4-j} = 0.69$$

$$Q(1) = \sum_{0}^{1} \binom{4}{j} 0.2^j (0.8)^{4-j} = 0.82$$

$$Q(2) = \sum_{0}^{0} \binom{4}{j} 0.1^j (0.9)^{4-j} = 0.66.$$

The probability of obtaining the required damage is greater for mix $k = 1$ than for any other mix. Thus, $k = 1$ is the optimum mix and yields a probability of success of 0.82.

In this chapter we have attempted to illuminate the factors linking the ECM equipment to the system and to the mission. We have also presented general techniques for forming the linkages in an optimum manner. These considerations are of importance not only to the ECM designer but also to the airframe designer and to those charged with the procurement of ECM and weapons as well.

In the first part of this chapter we dealt with optimization of ECM characteristics, subject to constraints imposed by the aircraft and to consideration of the available amount of knowledge of the enemy defense system. These boundary conditions are of primary concern in the design phase. In the second part we dealt with optimization of the distribution of aircraft payload between ECM and weapons, which is of primary concern in the procurement phase. However, since the optimum distribution tends to establish the desirable ECM equipment size, it has significant implications in establishing the constraints and therefore the design considerations.

It has been somewhat blithely assumed that defense effectiveness can be expressed analytically in terms of defense and ECM characteristics. In some cases it is possible to do so by theoretical considerations, but there is no satisfactory substitute for experimental data. Such data are difficult to come by but where they are available they should be employed as extensively as possible. It must be kept in mind, however, that the ECM picture is highly transitory, and that the value of data may be greater for validating theoretical influences than for the empirical evaluation of effectiveness.

Finally, it is well to realize that the relationship between ECM and defense effectiveness is so complex and is subject to so many uncertainties that any technique, analytical or otherwise, will provide only approximate solutions. Even if analysis is only a means of gaining insight and acquiring approximate guidelines, its contribution to increasing the effectiveness of an ECM effort can be significant.

8

SOME ASPECTS OF ELECTRONIC WARFARE IN THE SPACE ERA

INTRODUCTION

The conquest of space provides the newest dimensions for the conduct of warfare. Space applications of electronic warfare are therefore a fitting subject for this, the concluding chapter of the book.

In this chapter we shall discuss in a general way some of the possible applications. The realm of space combat still holds many unknowns and it is therefore too early to obtain complete solutions to many of the problems to be encountered. It will be our purpose here to point out some of the major factors that will influence the engineer in creating electronic weapon systems for use in space.

Specifically, we shall investigate some of the influences of "space" on weapon-system planning. A brief description of the environment will be presented, and will be followed by a discussion of some problems associated with the use of communication, reconnaissance, antennas, infrared, and radar systems from the point of view of electronic warfare.

The roles of space in warfare can be separated into three categories. The first, and most often discussed, is that in which space is employed to augment terrestrial warfare. Examples are: satellite warning systems, orbital bombardment systems, and satellite relay stations for terrestrial communications. The second category is that in which warfare is conducted between artificial elements in space and corresponds to combat among naval vessels in the ocean. The third is warfare involving planets other than Earth, or natural satellites. Attacks on enemy bases on the moon, war with the Martians or Venutians are examples. This text will deal primarily with the first category, that in which space is employed to augment terrestrial warfare.

There are two factors, both due to the Earth's characteristics, which give space the potentiality of augmenting terrestrial warfare. These are the Earth's curvature and gravity. Curvature limits the range between object and observer according to the expression:

$$R_{12} = K[\sqrt{h_1} + \sqrt{h_2}] \tag{8-1}$$

where R_{12} = maximum distance between object and observer in miles,

h_1 = altitude of observer in feet,

h_2 = altitude of object in feet,

K = constant dependent on the curvature of the Earth and the propagation characteristics. For radar, $K \approx 1.3$ miles/(ft)$^{1/2}$.

The important point indicated in Eq. 8–1 is that the observation range increases as the one-half power of the observer's altitude. From an altitude of 20,000 ft the range of observation of a point on the Earth's surface is 180 miles. At an altitude of 80,000 ft, which is near the limit of aerodynamic flight, the range is only twice this value. To obtain sightings at intercontinental range an altitude in excess of one thousand miles is required, which is well into the space domain. The vantage afforded by space is therefore a necessity for a practical global reconnaissance system. For similar reasons space provides a potential solution of the problem of reliable global communications.

Earth's gravity requires the expenditure of extensive amounts of energy to place an object in space, and although it is expensive to provide extraterrestrial means of augmenting terrestrial warfare, it is also expensive to attempt to destroy these means. In addition, the lack of a sensible atmosphere in space makes it a relatively simple task to provide decoys and thus to amplify the cost of destruction. As a result, one of the greatest problems associated with defense against the ICBM is decoy discrimination.

It is interesting to note that decoy considerations suggest a twofold advantage in the use of a satellite system to implement defense against the ICBM. First, a satellite can observe the ICBM during the phase of its flight prior to the time when decoys can be ejected. Second, the satellites can be decoyed readily to prevent their destruction.

A major requirement for a strategic deterrent system (a system intended to deter a nation from launching an all-out nuclear attack) is that the system be able to survive an initial enemy attack. Some of the alternatives suggested and employed to date to meet the requirement for invulnerability are dispersion, "hardening," warning coupled with a quick response, mobility, and the employment of orbiting or moon bases.

Dispersion serves to decrease the number of deterrent system elements destroyed by a single enemy weapon; "hardening" decreases the effective radius of the enemy weapon and thereby decreases the probability that the weapon will destroy an element; and mobility decreases the probability that an element will be at the intended point of aim of the weapon. As weapon-guidance technology improves, the influence of hardening decreases to that of dispersion, i.e., one element destroyed per enemy weapon. Therefore, if a nation relies on dispersion or hardening its deterrent force will be vulnerable to a surprise attack by an enemy with a force of only slightly greater size. A short warning period coupled with a quick response has two disadvantages: first, that the system may be employed by error; and second, that the time to make a decision to employ the system can easily exceed the warning time available. The chance of the system being employed by error is much more severe when the elements are missiles than when they are bombers, which can be recalled. However, future systems will consist mainly of missiles because they surpass bombers in penetration survival ability, and greater mobility. Thus, among the terrestrial means of providing invulnerability to the initial enemy attack, mobility provides the most promising answer.

Consider, now, the extraterrestrial means—satellite or moon bases. A system of orbiting elements, adequately decoyed, would require an extremely costly attack of anti-satellite weapons launched from Earth. However, the time is not too remote when manned vehicles capable of examining and disabling such a system will exist. Today such a capability does not exist, but neither does an orbiting deterrent system, and the time lapse between the appearance of each will be relatively short.

Moon basing of the elements of the deterrent system will be worthy of consideration as a means of providing invulnerability only to the nation that controls the moon. If both nations have bases on the moon they will be more vulnerable to surprise attack than similarly configured bases on Earth. This weakness will result from the shorter lunar distances and weaker gravity. By virtue of the long travel time, elements based on the moon will be relatively invulnerable to attack from Earth. However, supreme control of the moon by any one nation does not appear likely. The employment of extraterrestrial bases, either satellite or lunar, does not appear to offer a theoretically promising solution to adequate invulnerability for a strategic deterrent system. Though it is best to avoid prophesying, because of the pitfalls inherent in that art, it may be said that the primary roles of space in augmenting terrestrial warfare will include communications, reconnaissance, and possibly, defense against ICBM attack.

SOME ELEMENTS OF THE SPACE ENVIRONMENT

Two ensembles of electro-mechanical specifications are, in general, applicable to electronic equipment. Functional requirements of the particular mission must of course be satisfied. These ordinarily encompass such operational parameters as transmitted power, antenna gain(s), receiver sensitivity, channel capacity or band width, operating frequency, and so forth. An additional requirement, equally important, is the equipment's tolerance for environment. Survival specifications indicate the extremes of environmental factors to which the equipment may be subjected for a period of time without permanent loss of functional capability. It need not necessarily operate normally during exposure to these factors, but must do so after their removal. Specifications of functional environmental tolerance, however, indicate the conditions under which the equipment must be capable of satisfactorily meeting its performance specifications.

In the context of space operations, survival specifications are related primarily to ability to withstand the conditions of handling, shipping, storage, launch, atmospheric re-entry, and/or landing, when applicable. The actual characteristics of the space environment, on the other hand, are critical in defining the conditions under which the equipment must function properly. Factors related to temperature, radiation, prime power sources and total power consumption are typical of functional considerations upon which the space environment bears a direct influence. Size, weight, and shape restrictions are also imposed by the nature of each mission. Fortunately, certain characteristics of the space environment may offer at least partial solutions to these problems, which are presented primarily by the necessity for initial escape from Earth.

The functional and environmental requirements for equipment to be used in spacecraft are particularly stringent because of a combination of factors arising from launching conditions, the physical environment of extraterrestrial space, and the great distances involved in interplanetary space communications. The magnitude of the latter problem is best illustrated by a few simple calculations. The power actually reaching the receiving equipment of a communications link may be expressed as

$$P_r = \frac{P_t G_t G_r \lambda^2}{(4\pi R)^2}. \tag{8-2}$$

The corresponding minimum received power required for a specified probability of detection (or reliability of communication) may be written

$$P_m = kT_e B\left(\frac{S}{N}\right). \tag{8-3}$$

It will be instructive to take the ratio,

$$X = \frac{P_m}{P_r}, \tag{8-4}$$

and to substitute arbitrary (but somewhat practical) values of the various parameters, as a starting point. The numerical value of X will then be an indication of *required improvement*. Substituting Eqs. 8–2 and 8–3 into 8–4,

$$X = \frac{kT_e B\left(\dfrac{S}{N}\right)(4\pi R)^2}{P_t G_t G_r \lambda^2} \tag{8-5}$$

and initial sample values of the parameters are taken as follows:

k = (Boltzmann's constant) = 1.38×10^{-23} w/cps°K,

T_e = equivalent noise temperature, initially taken as 580°K for a receiving system at 17°C = 63°F = 290°K, with a noise figure of 3 db,

B = effective receiver band width, initially taken as 1 kcs,

S/N = signal-to-noise ratio required for satisfactory detection, typically 16 db,

R = communications range, taken here as 10^{11} meters for vicinity of Earth to vicinity of Mars at closest approach,

P_t = transmitted power, taken initially as 1 w,

G_t = transmitting-antenna gain, taken initially as 0 db,

G_r = receiving-antenna gain, taken initially as 0 db,

λ = wavelength, taken as 30 cm (frequency = 1 kmcs).

Substituting these initial values into Eq. 8–5,

$$\therefore\ X \approx 10^{10}.$$

Expressed in decibels, $X \approx 100$ db. That is, an overall improvement in system sensitivity of approximately 100 db is required for this interplanetary range, although the parameters given would be satisfactory for short ranges of about 10^6 meters, or approximately 500 nautical miles.

The only areas in which system improvements of many orders of magnitude (over those assumed above) may reasonably be expected, are those of transmitted power and of antenna gain(s). In the latter case, antenna gains ranging from a few db up to perhaps 40 db or more may be within reason. The weightlessness and essential absence of any significant gaseous atmosphere beyond a few Earth radii into space may permit the utilization of very large, lightweight (in terms of payload weight at launch) antenna structures erected after the vehicle reaches a space environment. High-gain antennas, with their narrow beams, impose the requirement for accurate antenna tracking, however. In passing through Earth's atmosphere at anything but normal incidence, electromagnetic waves are refracted. The extent to which the resultant antenna pointing error is significant depends, of course, on the beam width and other factors. Thus, the advantages of antenna gain must be compared with the weight, complexity, reliability, primary power, and other considerations associated with antenna tracking and stabilization. The result may be used to determine an optimum trade-off between additional antenna gain and increased transmitter power and/or receiver sensitivity. In the case of a simple, unstabilized satellite, for example, it may be more economical to increase system sensitivity a few db by employing low-noise receiving techniques than to exchange directive for nondirective antennas aboard the satellite.

Selection of a communications frequency or frequencies is of great significance from two standpoints. The aperture of an antenna used for receiving depends primarily upon its physical size and is, to a first approximation, independent of frequency; the antenna's gain, however, is inversely proportional to wavelength squared. This relationship, as discussed in Chapter 6, is:

$$G = \frac{4\pi A_{\text{eff}}}{\lambda^2};\tag{8–6}$$

A_{eff} is the effective antenna aperture and λ, the wavelength, is, of course, inversely proportional to frequency. These considerations in the choice of optimum communications frequency is not unique with space applications, although the lack of atmospheric absorption problems may, in pure space applications, simplify the problem somewhat. Frequency selection is influenced, however, by considerations of solar and cosmic noise, and by terrestrial noise when one end of the circuit must operate in or near Earth's atmosphere.

The equivalent noise temperature, T_e in Eq. 8–5, might be improved in some cases by exploiting a characteristic of the space environment that results from its lack of atmosphere. Since essentially no gaseous atmosphere exists to diffuse and absorb solar radiations, extremely dark shadows occur behind opaque bodies. Temperatures of only a few degrees K, which exist in shaded areas, might be utilized to allow use of maser amplifiers or other low-temperature, low-noise devices without the need for complex cryogenic systems. T_e consists of an effective sky temperature plus an effective receiver-noise temperature. The latter may be reduced to the vicinity of 10 degrees K with modern low-temperature techniques; there is no point in lowering it much below the effective sky temperature, which is a function of cosmic noise (and terrestrial noise, in the case of locations within Earth's atmosphere). Figure 8–1* shows the approximate equivalent sky-temperature range associated with cosmic and terrestrial noise. It is apparent that, for a system operating entirely outside Earth's atmosphere (no terrestrial noise), frequencies of 1 kmc or more are desirable in order to make optimum use of low-noise receiving techniques. If one end of the system lies within the atmosphere, so that terrestrial noise becomes significant, it is obvious from Fig. 8–1 that the upper useable frequency limit is considerably restricted. In any case, antenna gain for a given physical size does not increase indefinitely with increasing

* D. C. Hogg, "Effective Antenna Temperature Due to Oxygen and Water Vapor in the Atmosphere," *Journal of Applied Physics*, **30**, 9, Sept. 1959, pp. 1417–1419.

frequency. A limiting point is reached, above which equipment state-of-the-art, fabrication and stabilization techniques, and possibly other factors not yet recognized, neutralize the potential advantages of going to even higher frequencies. For an extraterrestrial system, T_e may be limited to perhaps 10 degrees K, yielding a reduction in X (Eq. 8–5) of about 17 db. Choosing a frequency of approximately 3 to 5 kmcs allows antenna gains of the order of 25 db for both transmitting and receiving, without introducing undue size

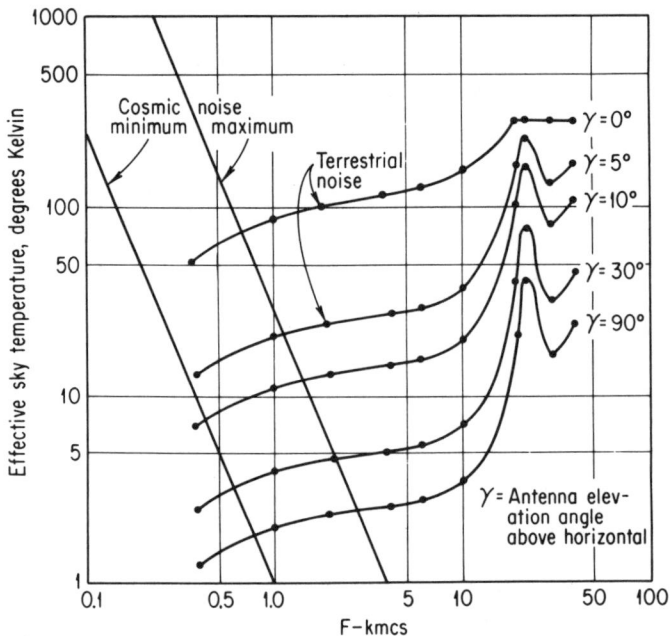

Figure 8-1. *Effective sky temperature due to cosmic and terrestrial noise sources vs. frequency.*

or stabilization problems. The total reduction in X thus obtainable from noise-reduction and antenna gain is of the order of

$$\Delta X \approx -17 - 25 - 25 = -67 \text{ db}.$$

The remaining improvement required must be furnished in the form of increased transmitter power or decreased system band width. If the channel is to be useful for simple voice communications, for example, a minimum band width of approximately 2 kcs (which yields $\Delta X = +3$ db) and consequently an increase in P_t of $(100 - 67 + 3) = 36$ db, to 400 w, are required.

In practice, the determination of system configuration is not quite as straightforward as might seem to be implied above. In particular, it should be obvious that interplanetary communications, because of their heavy reliance on low-noise receiving conditions, are highly susceptible to countermeasures. Noise jamming will, in effect, raise the "sky temperature," hence increasing T_e. The utilization of high-gain antennas with selective, narrow beams is made yet more attractive by this consideration. The value of utilizing extremely low-noise receiving equipment is critically dependent upon reduction of noise picked up by the antenna. In some cases (e.g., Earth satellite-to-moon communications link), it may even be desirable to choose an operating frequency subject to heavy atmospheric absorption, so that jamming from Earth is rendered impractical.

In the preceding discussion we have provided some appreciation of the functional problems of employing electronic systems over distances of the magnitude involved in space operations. Although optical paths and comparatively low noise temperatures provide near-optimum conditions, the great distances involved nevertheless lead to technological problems.

The actual physical environment encountered outside Earth's atmosphere imposes additional constraints on the design of electronic equipment for spacecraft. Temperature extremes, ranging from a few degrees K in shadows to several hundred degrees in areas of direct solar illumination (see Fig. 8–2)* certainly impose the requirement for some form of artificial temperature stabilization within a spacecraft. The degree to which such stabilization is accomplished determines, of course, the degree to which electronic equipment must be immunized to the effects of ambient temperature. A number of temperature-stabilization techniques have been used or proposed; both passive systems, such as patterns of reflective and absorptive materials coated on the surfaces of space vehicles, and active systems, involving artificial heating or refrigerating systems, are practicable.

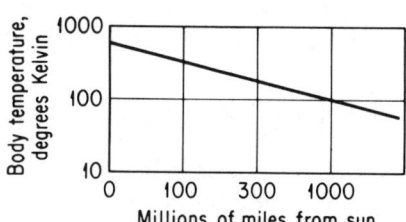

Figure 8-2. *Equilibrium temperature of a black body in space as a function of distance from the sun.*

The exact system that will prove most desirable in a given case depends upon the specific nature of the vehicle and its mission. Advances in the field of temperature stabilization have been accompanied by development of electronic circuitry techniques suitable for use under a wide range of temperature conditions. It is, therefore, to be expected that temperature problems will not be major determining factors in future space flight.

More serious consequences are projected by other forms of electromagnetic and particle radiation, however. It has been estimated that a meteoroid of visual magnitude 17 will penetrate approximately 0.005 inch of aluminum, whereas one of magnitude 5 will penetrate roughly two inches. From Fig. 8–3, it is found that approximately one meteoroid of magnitude 17 will be encountered per day by a sphere of 3 meter diameter located near the Earth

* From H. A. Manoogian, "The Challenge of Space," *Electronics,* **32,** 17, April 24, 1959.

but outside its atmosphere.* A meteoroid of magnitude 5, which of course constitutes a much more serious problem of protection, will be encountered only once in about 135 years. For any given space mission, the difficulty of implementing "armor" protection must be balanced against the risk projected by meteoroids. Although the danger of meteoroid impact is one requiring attention to mechanical design and protection rather than to electronic specifications, the question of antenna and radome design may enter in, especially when very high reliability is required, as in the case of manned spacecraft.

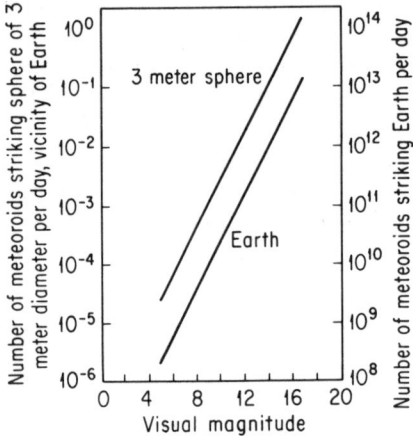

Figure 8-3. *Visual magnitude of meteoroids.*

Two areas of high-energy electron radiation were discovered in early Earth-satellite probes in 1958, and are often referred to as *Van Allen radiation belts*, after one of the principals in the work that led to their discovery and explanation. The dense portions of these belts appear to extend from altitudes of approximately 500 or 600 miles to perhaps 20 or 30 thousand miles above Earth's surface, with peak radiation densities equivalent to between 10 and 100 roentgens per hour.† This radiation density is far in excess of that to which a human being may be safely exposed, even for a short period. In addition, this and other forms of nuclear radiation encountered in space have serious implications in so far as solid-state devices are concerned. The operating parameters of transistors, for example, have been found to be seriously, and to a large extent permanently, degraded by bombardment by fast particles, including neutrons and electrons. Tunnel diodes, on the other hand, are considerably less susceptible to "radiation poisoning," and vacuum tubes are almost impervious. The radiation threat may restrict the designer's choice of components for space electronics, since little margin can be allowed for reduction (or elimination) of equipment performance (e.g., communications range) due to deterioration of components. The electronics problem may be somewhat simplified in the case of manned vehicles, wherein protection from excessive radiation must be provided for the human occupants. Consideration must also be given to the choice of launching paths (or geocentric orbits, in the case of satellites of Earth) to take advantage of the essential absence of the high-energy electrons in polar regions. Van Allen

---

* From Buchheim, et al., "Some Aspects of Astronautics," IRE Transactions on Military Electronics, **2**, 1, December 1958, p. 16.

† J. A. Van Allen, "Radiation Belts Around the Earth," *Scientific American*, **200**, 3 March 1959, pp. 39–47.

radiation is thought to result from the trapping of solar or cosmic electrons by Earth's magnetic field, whence individual electrons spiral along magnetic lines of force from polar region to polar region, until their energy is consumed by collisions with other particles. A number of the other planets possess magnetic fields not unlike that of Earth, and even more severe high-energy radiation belts may exist in their vicinity. Thus the development of vehicles and systems capable of safely traversing such regions is important, even though departure from Earth may be accomplished via low-radiation polar routes (see Fig. 8–3(A)).

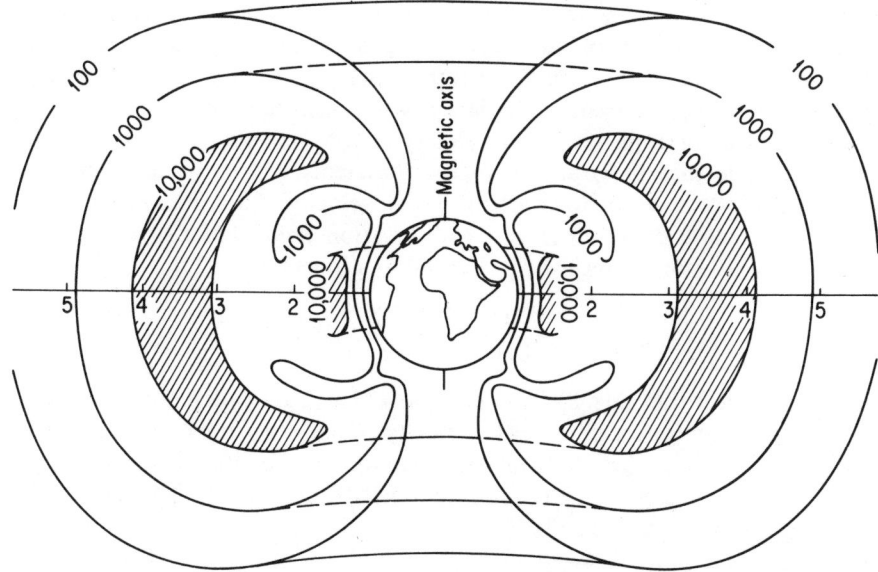

Figure 8-3(A). *Relative "Van Allen" radiation vs. distance from Earth (in Earth radii). From: J. A. Van Allen, "Radiation Belts Around the Earth",* Scientific American, **200**, *No. 3, March* 1959, 39–47.

In addition to the known sources of radiation, including high-energy charged particles and gamma rays ("hard" X-rays), there exists a strong probability that hitherto undetected particle or radiation phenomena will be discovered in the course of extended space exploration.

The great distances involved in interplanetary operations introduce an additional problem, in that transit time for electromagnetic waves can no longer be considered unimportant. Signals travelling at the speed of light, 3×10^8 meters per sec, require approximately 1.27 sec to reach the vicinity of the moon from Earth, about 8 minutes average for Mercury or Venus, (varying from 14 minutes to 2 minutes in the latter case, depending upon orbital position), and about 5 hours to the edge of the solar system. The problems of two-way communications are obvious. Since the great distances also demand great vehicle velocities if travel times are to be kept within reasonable limits, relativistic effects and comparatively large Doppler shifts enter into the problem. For a vehicle moving radially at a fraction, Y, of the velocity of light (c), the Doppler-shifted frequency, $f(Y)$, as a function of Y is

$$f(Y) = \sqrt{\frac{1+Y}{1-Y}} f(0), \tag{8-7}$$

where $f(0)$ is the unshifted nominal frequency. The implications of Eq. 8-7 for narrow-band systems are obvious. Some means must be provided at the receiving end of each line for either continuously tracking the received signal frequency from the time of launch, searching in frequency in order to find and lock onto the signal, or furnishing transmitter and receiver with extremely precise frequency standards. The latter would also require highly accurate control and knowledge of relative velocity in order to predict Doppler shift; it is probable that the tracking types of systems will enjoy predominant popularity for some time to come. The high velocity of a space vehicle moving in a receding direction from an observation point will also result in a relativistic reduction in received power, P_r, as a function of Y, thus:

$$\frac{P_r(Y)}{P_r(0)} = \frac{(1+Y)^2}{1-Y}. \tag{8-8}$$

It should be noted in Eqs. 8-7 and 8-8 above that Y is positive for approaching and negative for receding radial velocity. Consider, for example, a transmitter on a vehicle receding from the Earth at a rate of 10^4 meters/sec. The transmitter frequency is 1 kmcs and its power output 1 kw. The frequency received on Earth is, from Eq. 7-7:

$$f(Y) = \sqrt{\frac{1+(-10^4/3 \times 10^8)}{1-(-10^4/3 \times 10^8)}} \times 10^9 = 0.999967 \times 10^9 \text{ cps}$$

or
$$f(Y) = 999.967 \text{ mcs}.$$

Thus the received frequency, $f(Y)$, has been shifted downward by about 33 kcs from its nominal value, $f(0) = 1000$ mcs. From Eq. 8-8,

$$\frac{P_r(Y)}{P_r(0)} = \frac{\left[1 + \left(\frac{-10^4}{3 \times 10^8}\right)\right]^2}{1 - \left(\frac{-10^4}{3 \times 10^8}\right)} \doteq 1 - 10^{-4} \text{ or } 0.9999.$$

The received *power* is reduced in this case by only 0.01 % because of relativistic effects. An attenuation of 10 % would result from vehicle velocities of about 10^7 meters/sec, or $\frac{1}{30}$ the velocity of light.

A number of methods have been developed or are under development for generation of prime electrical power for space operations. Various forms of chemical batteries have been used successfully to provide modest amounts of power for early unmanned space missions. Weight and lifetime limitations make batteries appear most useful as storage and stand-by sources for use in conjunction with other prime power sources. Solar energy has been utilized directly in the production of comparatively small amounts of electrical power. Although the lifetime and reliability of silicon crystal solar cells are high, their utilization will probably continue to be restricted to comparatively low-power applications, because of their relatively low power-to-weight ratio and large physical area. Solar cells, used with chemical batteries as storage media to provide power during periods when direct solar illumination is not present, will probably continue to offer one of the most simple and reliable solutions to the power problem for small satellites.

The development of electrical propulsion systems for deep-space missions will bring with it vastly increased demands for primary electrical power. The thermionic converter, which converts intense heat into electrical energy, is one developmental device that offers the probability of higher power-to-weight ratios than the devices mentioned above. It, too, is limited by the amount of incident solar power, which is of the order of 1 kw/sq meter at 10^8 miles from the sun. Since this density decreases as the square of distance from the sun, the practicality of any type of solar-power converter becomes more questionable as deeper probes into interplanetary (and eventually interstellar) space come under consideration. It seems highly probable that nuclear power sources, either isotope-powered or based on the use of some type of reactor, will carry the brunt of electrical (and mechanical) power demands in large-scale space operations of the future. As attempts to generate electrical power directly from the fission process become successful, the result should be an extremely practical, economical, and long-life source of great quantities of power for communications, propulsion, and functional uses within the space vehicle.

An additional, intriguing idea has been repeatedly suggested as a means of capitalizing on the near-perfect vacuum existing in space. High-quality vacuum tubes might be constructed without envelopes and their attendant weight and bulk.

In the remaining sections we shall consider the objectives of specific electronic systems in the context of the space environment.

SPACE MISSIONS—COMMUNICATIONS AND RECONNAISSANCE

The conquest of space made possible many operations, both military and commercial, that were previously impossible. One of the most formidable and best publicized is the delivery of the Intercontinental Ballistic Missile (ICBM). Running a close second in recognized importance is the field of global communications.

The problem of global communications has long been a major concern to all of our armed forces. The Army must control combat forces in remote areas, the Navy constantly has ships at sea, and the Air Force must maintain contact with Strategic Air Command bombers in flight. Each of these services has undertaken to solve its particular communications problem with the techniques and technology at hand. Unfortunately, in all cases the results left something to be desired.

The problem reduces quickly to one of frequency spectrum characteristics. The high-frequency portion (3–30 megacycles), although capable of covering the ranges involved, is unreliable because of atmospheric noises and its high degree of dependence on the selection of the optimum operating frequency. The VHF portion (30–300 megacycles), although reliable enough, is limited to useable ranges of 300–400 miles before a relay station is required.

A possible solution for the military needs is foreseen through the use of some form of a global satellite communications system if the obstacles of ECM can be overcome. Such a system will allow the use of higher frequencies providing sufficient band width for large quantities of data, and is relatively independent of problematic propagation phenomena associated with the unreliable lower frequencies.

There are two basic modes of operation for Global Communications Satellite Systems (GCSS). The first method involves the use of active satellites; that is, satellites carrying sufficient equipment to receive, *amplify*, and retransmit the signal. The second method would use passive satellites. This form of satellite serves only as a reflector from which the signal is "bounced" to the desired receiving point. Figure 8–4 shows the basic difference between these two systems. The number of satellites used in the constellation and the height of their orbits are closely related to the mode of operation chosen, i.e., passive or active.

The merits of a satellite system in providing the military with a global communications capability are well recognized. However, it is not a clear decision as to whether a passive or an active system is the most desirable. Discussions of this problem that have appeared in the unclassified literature are generally concerned with commercial systems and neglect the influence that electronic-warfare techniques have in degrading system performance in arriving at their conclusions.

It will be worth while here to examine some of the advantages and disadvantages of each mode and to consider aspects of the systems of importance in the hostile electronic environment that might result in wartime. The techniques of electronic warfare could be directed very profitably against a global communications system upon which much of a country's military services are dependent. It is therefore reasonable to consider systems in the presence of possible jamming methods.

The variables associated with the satellite system configuration are many and include such fundamental considerations as the total number of satellites

SPACE MISSIONS 189

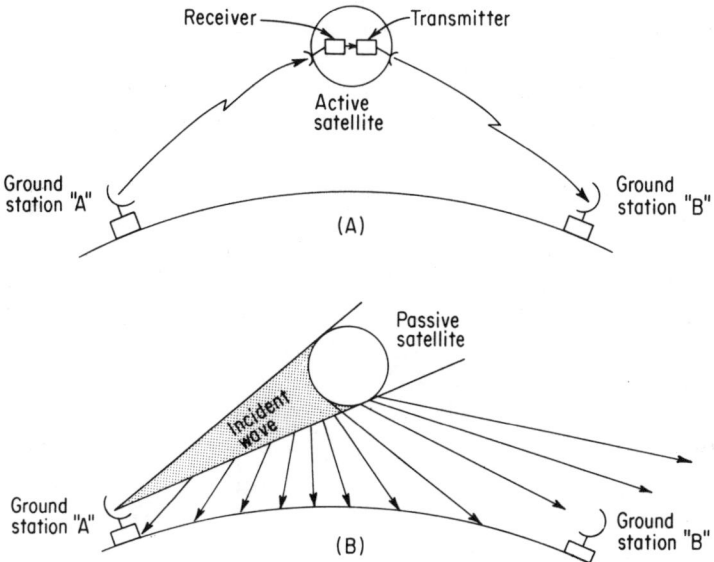

Figure 8-4. *Comparison of basic operating principles for active and passive satellite communication systems.*

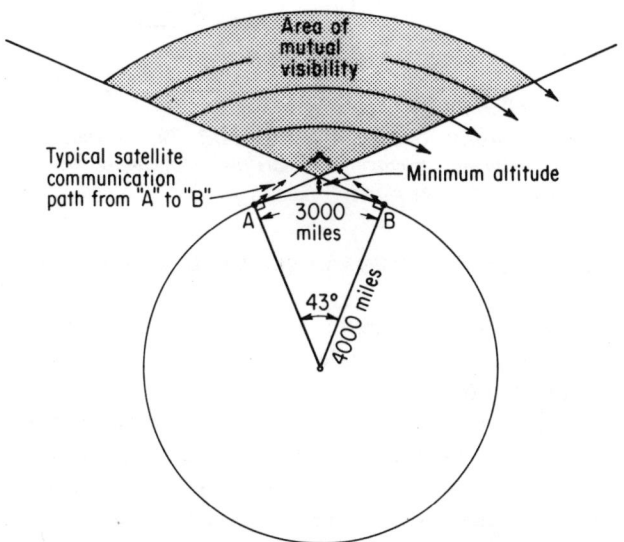

Figure 8-5. *Typical configuration for satellite communication link.*

involved and their orbital altitude. To reduce the problem to a manageable degree it is advantageous to define a mission objective and then discuss only systems that could logically be considered in accomplishing the defined mission. For this purpose the objective is to provide communications between two points on the Earth located 3000 miles apart. Also, let these points lie on a great-circle path in the plane of the satellite orbit on a non-rotating circular Earth.

The fact that the two stations, A and B (see Fig. 8-5), are 3000 miles apart establishes the minimum possible satellite altitude as 300 miles. However, this placement would allow line-of-sight paths from station A to the satellite, to B only when the satellite is exactly at the midpoint. This arrangement is, of course, impractical, hence a higher altitude is required. The upper altitude

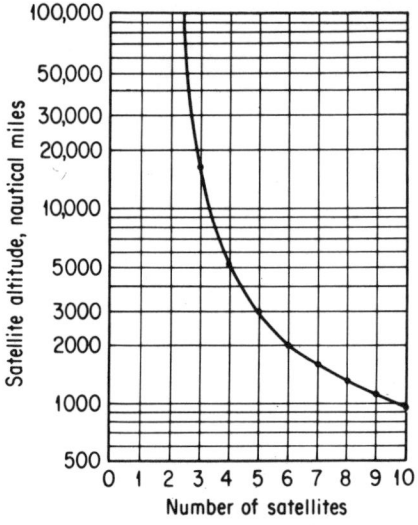

Figure 8-6. *Number of equi-spaced satellites required to provide continuous coverage between points A and B located as described in text.*

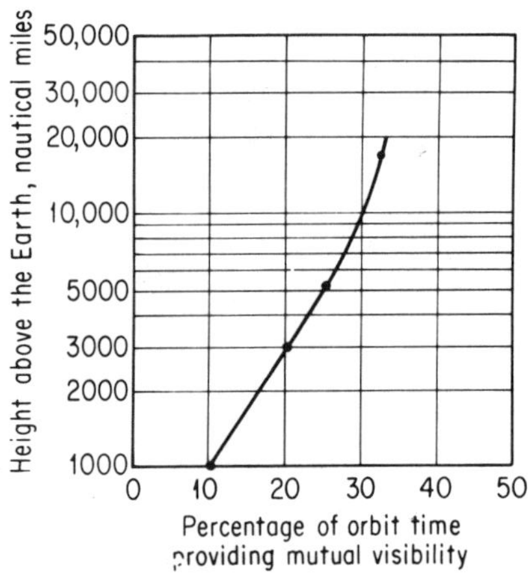

Figure 8-7. *The percentage of time a single satellite is mutually visible from points A and B.*

is limited by the maximum ranges of present transmitting and receiving equipment.

The number of satellites required is a function of how they are placed in their orbit as well as of their altitude. They may be either equi-spaced or randomly spaced within their orbit.

With equal spacing it is possible to establish the conditions for continuous communications. Figure 8-6 shows a plot of the number of satellites required vs. orbital altitude to maintain communications between points A and B. The possibility of satellite-to-satellite relaying is not being considered here.

With random spacing* within the orbit it is necessary to consider the

* See Pierce and Kompfner, "Transoceanic Communications by Means of Satellites," *Proceedings of the IRE,* March 3, 1959, p. 376.

probability that the satellites may so space themselves that a time will occur when none are visible simultaneously from points A and B. Any such system designed for either commercial or military requirements must provide a reliability of at least 90 per cent or higher. If the desired probability that a satellite can be seen simultaneously from A and B is P, then the probability that it will not be seen is $1 - P$. The fraction of its orbit during which it is visible, V, is a function of its altitude, and the fraction of time it is not visible is $1 - V$. The percentage of time a single satellite is mutually visible from A and B as a function of altitude is shown in Fig. 8–7.

If N satellites are used in the constellation it is now possible to write:

$$1 - P = (1 - V)^N \tag{8-9}$$

$$N = \frac{\ln(1 - P)}{\ln(1 - V)}. \tag{8-10}$$

N is plotted in Fig. 8–8 for two system-reliability probabilities as a function of altitude.

Hence, for a 2000-mile-high orbit the percentage of visibility (determined from Fig. 8–7) is 16%. If the desired reliability of sighting is $P = 90\%$, Eq. 8–10 is:

$$(1 - 0.9) = (1 - 0.16)^N$$

or

$$N = \frac{\ln(0.1)}{\ln(0.84)} = \frac{-2.3}{-0.17} = 13.5 \approx 14.$$

Therefore, 14 satellites are required to accomplish the mission. It is this function that is plotted in Fig. 8–8.

So far, we have been concerned with defining possible satellite configurations. Nothing has been said as to whether these are active or passive satellites, what the ground station characteristics are, or where potential jamming methods might be employed.

Consider, now, that a satellite is at the midway point between A and B at an orbital altitude of 3000 miles. From Fig. 8–6, five satellites would be used in the constellation if they are equally spaced, and from Fig. 8–8, a minimum of eleven if they are randomly spaced. Figure 8–9 shows the satellite position in relation to A and B.

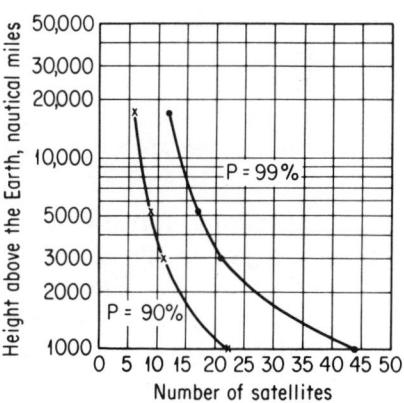

Figure 8-8. *Number of randomly spaced satellites for two probabilities of mutual sighting from points A and B.*

Assume, now, that the satellite must pass over unfriendly territory and hence a jammer is located directly beneath the satellite at point C in Fig. 8–9. If the satellite is a passive sphere it is obvious that it will scatter the incident plane wave received from Station A uniformly in all directions. The echo area σ for a reflecting sphere is simply πr^2, where r is its radius. It can therefore be seen that the monitoring receiver associated with the jammer at C will have no trouble receiving A's transmitted signal. This signal will, of course, provide all of the necessary information to initiate jamming action if desired. However, a passive satellite is not limited in either band width or operating frequency as is an active satellite. This allows the link between A and B to rapidly change frequency as a powerful ECCM technique. It is not our purpose here to delve into a detailed analysis of path losses and system sensitivities, but rather to cover in general terms the relative factors of ECM applications in space. Therefore it will be stated that such a link as described here is quite technically possible* with present equipment and a passive satellite in the order of 100 ft in diameter.† The path from A to satellite, to B is approximately 7200 miles. For the condition shown, the path from jammer to satellite to B is 6600 miles. This small difference in distance (unfortunately in favor of the jammer), is of minor importance and is only cited to show that if communications can be established over the path outlined they can also be jammed from the point indicated by a transmitter of approximately equal power.

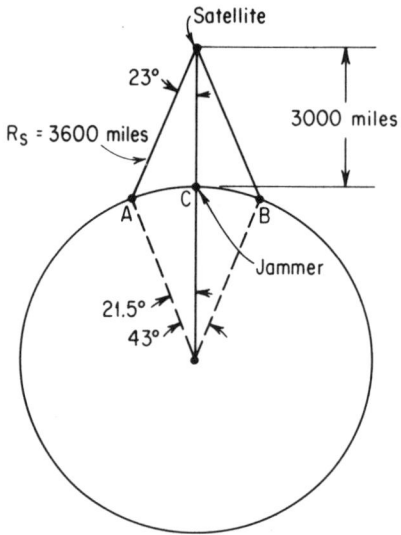

Figure 8-9. *Configuration for a satellite located midway between points A and B with a jammer at point C.*

Now, consider Fig. 8–9 with all positions held the same and the satellite shown to be an active element in this case. This situation requires that primary power be available in the satellite. If the antenna system used on the satellite is of the omnidirectional type, the situation does not change much from the case discussed earlier, except that the receiver being jammed is now in the satellite rather than at point B. On the other hand, with primary power in the satellite it is now possible to use a steerable antenna at that point. The power received by the satellite receiver from the jamming transmitter is given by:

$$P_{r_j} = \frac{P_j G_j G_{sl} \lambda^2}{(4\pi R_o)^2}. \tag{8-11}$$

* Pierce and Kompfner, "Transoceanic Communication by Means of Satellites," *Proceedings of the IRE*, March 3, 1959.

† Since the preparation of this manuscript such a system has been successfully tested using a sphere 100 ft in diameter. The system is known as project ECHO.

R_o is the range from the jammer to the satellite and P_j and G_j are the power and antenna gain of the jammer system. G_{sl} is the satellite antenna gain in the direction of the jammer. In Fig. 8-9 it can be seen that the time being considered represents an off-axis angle of 23.0 degrees to the jammer. This angle will, of course, vary with satellite position. Depending on the type of antenna system and the frequency being used (effective antenna aperture), this off-axis loss can represent a 20 to 30 db attenuation in the jamming signal. Since the distances involved in this case are about equal, the effective radiated power of the overall jammer system must be about 300 times greater than the transmitter system located at point A in order to have the same signal strength at the satellite receiver.

It should also be pointed out that the reliability of active satellites will leave something to be desired for some time to come and hence is another consideration when selecting a system to meet military requirements.

From this brief discussion an interesting conclusion is indicated. Satellite communications systems are inherently capable of providing contact over paths that no other present radio system can accomplish. This ability results from the very nature of the altitude of satellite orbits. However, there seems to be nothing inherent in the satellite system that provides immunity from jamming, although typical methods of frequency diversity and signal coding most certainly can be employed and should prove helpful.

It has been shown here that neither of the two system modes discussed above is inherently free from the evils of ECM, and that the engineer capable of designing a satellite system to meet military needs, that is, impervious to present jamming techniques, will indeed perform a worthwhile service.

A second interesting mission that is particularly adaptable to satellite systems is the function of reconnaissance. For this work satellites offer the distinct advantage of being able to cover large areas quickly, operation in a "free air space," being able to take many consecutive readings of the target and of operating from a desirable vantage point.

An approximate expression for the velocity of a satellite in orbit as a function of altitude is given by:

$$V_s = \sqrt{\frac{gR^2}{(R + \lambda)}}, \tag{8-12}$$

where V_s = orbital velocity of satellite, fps,

R = earth radius, feet,

g = gravitational acceleration of 32 ft/sec^2,

λ = orbital height above Earth's surface, feet.

A plot of this equation is shown in Fig. 8-10. This information is useful, since from it the length of time that a specific point on the Earth can be observed from the reconnaissance satellite, can be calculated as a function of altitude. This observation time is also a function of the location of the target point in relation to the plane of the satellite's orbit. That is, the point of interest may not necessarily lie directly beneath the satellite on a given pass, which fact will tend to reduce the allowable observation time on that particular orbit.

Since orbits below about 250 nautical miles experience a serious drag from the earth's atmosphere, and hence a greatly reduced lifetime, most missions must be accomplished from altitudes of 300 nautical miles or greater. This great height imposes some serious problems on the detector being used for reconnaissance in the satellite. Some of the obvious modes of detection, depending on the target, include photographic optical systems, infrared cells, radar, ferret receivers, and television. The possibility also exists that radiation detectors would be useful as a method for providing surveillance of thermonuclear bomb testing in remote areas of the Earth.

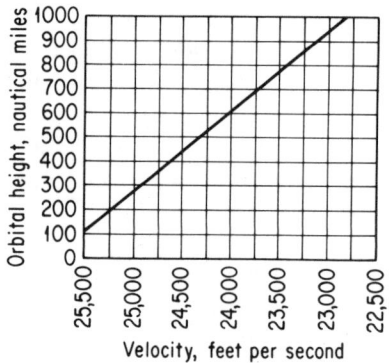

Figure 8-10. *Satellite velocity vs. orbital height, see Equation 8-12.*

If the satellite is used to observe the Earth beneath its orbit, the atmosphere will provide some degree of attenuation to the various detectors. Cloud cover will also inhibit the photographical/optical, infrared, and television systems on occasion. Nevertheless, these systems will prove effective information-gathering techniques. One of the major reasons why these systems prove so attractive for this purpose is the difficulty of "jamming" them. As far as is known, there is no active jamming method for optical systems. A brilliant light source (or infrared source) to completely obscure the background characteristics would represent a typical active jamming signal. However, passive decoys intended to cause optical confusion were developed and used in World War II (e.g., mock wooden aircraft to mislead bombers). Perhaps wooden ICBM launching sites will be constructed to provide such false information to reconnaissance satellites.

Possibly the most vulnerable element of a reconnaissance satellite is the data link used to transmit the information to the ground station. Fortunately the satellite can be made to store its data until it is over friendly territory; it can then be interrogated by a coded signal and will transmit its information to the ground station. The design must make this link secure so that the satellite cannot be interrogated as it passes beyond the enemy boundaries, making it "empty" when it arrives over the friendly site.

ANTENNAS IN SPACE

The application of satellites and other space vehicles for electronic-warfare purposes involves many antenna functions. There must be receiving antennas on the satellite to receive guidance instructions, command signals for the initiation or termination of self-contained operating equipment, messages to be re-transmitted to another satellite or to ground-based stations, and last and most important for reconnaissance purposes, signals from enemy emitters. Transmitting functions required of satellite antennas may involve the re-transmission of collected data or of messages when the satellite is a link in a communications chain. If an active radar system is carried by the satellite there must also be a transmitting antenna for the radar. Finally, the satellite itself may be designed as a passive radio reflector, such as a sphere or corner reflector, from which signals may be reflected from one ground station to another.

These diverse functions may be performed by many types of antennas, depending upon the radio frequencies involved, the band width required, and the power-handling capability needed. The weight and volume of the satellite that may be allocated to antennas will most certainly be held to the very minimum, and yet the reliability of the antennas must be the highest attainable. If an antenna, or for that matter any equipment, fails while the satellite is in orbit, maintenance is next to impossible.

The two principal categories or types of antennas commonly considered for satellite applications are flush antennas, and inflatable or unfolding antennas.

FLUSH ANTENNAS

Flush antennas must be used when operation is required during the launch phase and at altitudes at which the atmosphere is sufficiently dense to exert aerodynamic forces or heating effects on protruding antennas. Flush antennas may be simple slots with dielectric covers to provide low gain but wide angular coverage, or slots arranged in broadside or end-fire arrays to produce narrow beams if desired. For broader-band operation, spiral antennas such as equiangular or logarithmic spirals may be flush mounted on dielectric sheets backed by loaded cavities. At microwave frequencies very satisfactory high-gain end-fire arrays may be built almost completely flush by the use of surface or retarded-wave antennas. These antennas consist of corrugations in the surface of the vehicle fed by a low-profile horn or array. The beam width in the plane of the corrugated surface is essentially the beam width of the feed structure, but the beam shape in the transverse plane is modified by the corrugated slow-wave structure into a directive near end-fire beam. At lower frequencies it may be possible to insulate one part of the satellite body from another and to feed the two insulated sections as a dipole. This type of antenna is analogous to an insulated tail cap or wing tip used as a high-frequency antenna on a jet aircraft.

INFLATABLE AND UNFOLDING ANTENNAS

If antenna operation during transit of the lower atmosphere is not required, then aerodynamic effects become less important and the antenna structures on the vehicle may assume more conventional shapes and occupy a larger volume external to the body of the satellite after orbiting altitude is attained. In some cases it may be sufficient to have the antenna covered by a protecting shroud during the launch phase. After stable orbit is achieved the shroud may be blown away, leaving the antenna exposed. In other situations it may be desirable to erect the antenna from the satellite. A most common method familiar to all is the use of long flexible dipoles, as seen on the Russian Sputnik I. Such antennas may be coiled up or wrapped around the satellite while it is within the nose cone, and will spring out into place when the protective nose cone is blown off. Many spring-loaded antenna configurations can be devised. The helical antenna shape lends itself perfectly to such an operation. The helix is simply compressed flat and springs back into place when a trigger or cover is released.

The use of compressed-gas cartridges to drive antennas into place may also be desirable. In addition, the gas may be used to inflate metal-coated plastic film structures for use as antennas. Many conventional antenna forms such as horns, corner reflectors, parabolas, and spherical reflectors may be constructed of such film, folded within the satellite, and inflated after reaching orbiting altitude. The presence of micrometeorites that might deform such light structures must be considered. However, no internal pressure is needed to maintain the shape of these inflatable antennas when used on satellites. Since there is no gravitational force or aerodynamic drag, there is no tendency to deform the antenna structure. Figure 8–3 shows a relationship for meteorite impact on such an antenna.

POLARIZATION

Special consideration must be given to the selection of optimum polarizations to be used for antennas in space. It is axiomatic that no matter what polarization an antenna is designed to receive, there is always an orthogonal polarization that the antenna will not receive. Further, there are an infinite number of polarizations that the antenna will receive with varying degrees of effectiveness. Thus, a right circular polarized antenna will not receive left circular polarization at all, but will receive all linear polarizations with a 3 db loss. Conversely, a linearly polarized antenna will reject an orthogonal linear polarized signal and will receive both right- and left-hand circularly polarized waves with a 3 db loss.

As an additional example of the care that must be given to the choice of polarization in space, consider an antenna on a satellite that is stabilized so that the polarization is linear and parallel to the direction in which the satellite is moving. Furthermore, let it be assumed that the satellite is directly over Omaha, Nebraska, traveling from north to south. To an observer in San Francisco, California or Washington, D.C. the signals from the satellite antenna would be horizontally polarized, but to the observers directly under the orbit (north or south of Omaha) the signals will be vertically polarized. Each particular application of antennas to space vehicles must be analyzed specifically to determine the optimum polarization to be used.

ANTENNA MATERIALS FOR SPACE SYSTEMS

The environmental effects of space on antenna materials are many and varied. Much is yet to be learned about the behavior of materials in this environment, but some conclusions are already evident.

Materials with low vapor pressures must be used because of the very low ambient pressure experienced at the altitudes at which satellites are used. This consideration is particularly important when lubricants are required, since conventional oils and greases will soon evaporate. Graphite and molybdenum disulfide lubricants are the most satisfactory lubricants at present. Tetrafluoroethylene (Teflon) bearings, which are very satisfactory for highly corrosive environmental applications on Earth, are not suitable for space applications, not only because of low vapor pressure, but also because of radiation damage. Dielectric materials in general are very susceptible to radiation damage. Of common plastic dielectric materials Polystyrene is one of the most resistant to such damage. Other useable plastics include Mylar, polyethylene, and epoxy resins. Ceramics are generally very high in resistance to radiation damage as well as to high temperatures, and will be used more and more for space applications.

INFRARED APPLICATIONS

One of the most revolutionary developments of the last decade has been the growth and application of infrared (3×10^{-2} cm to 10^{-5} cm) technology to electronic warfare. The extreme precision and reliability of infrared seekers was demonstrated in a most unusual manner during the early testing phase of one missile program. The infrared guidance system was so accurate that direct hits were made on the engines of drone targets with such frequency that the cost of replacing the drones became excessive. Flares had to be attached to the extreme wing tips of the drone targets to divert the missile from the engine. (A direct hit on the wing tip would not be as likely to prevent recovery of the drone and its subsequent repair and re-use.)

Another equally famous application of infrared technology to warfare was the "Sniperscope," first used in the field during World War II. It enabled the infantryman to locate and shoot at his target under cover of complete darkness by detecting the infrared radiation from "warm bodies."

The absorption and scattering of the atmosphere limits the range of infrared systems, such as those mentioned above, to a relatively short distance. As soon as the infrared sensor is carried to high altitudes above the Earth's atmosphere in a satellite, the "atmospheric blindfold" is removed and greater effectiveness of infrared as a military tool is possible. As a celestial navigator, an infrared tracker has a unique advantage in the ability to select and discriminate between many stars and planets on the basis of their temperature. From its satellite-borne space platform an infrared system may also detect and track other artificial Earth satellites and ballistic missiles. It is quite likely that a satellite-borne infrared detector could give the earliest possible notice of the launching of an enemy ICBM by observing the great heat radiated by the rocket engines during launch.

The prospects of continuous reconnaissance and recording of the cloud cover of the Earth from a satellite are most promising, and long-range meteorological forecasting will benefit tremendously therefrom.

The integration of infrared and microwave radar into combined systems for precision-tracking and fire-control applications is a natural development. Reflecting optical systems can be designed to carry infrared and microwave detectors simultaneously. The microwave radar will measure the range to a target, whereas the infrared system with its much finer angular discrimination capability is used for precision angular tracking.

From the point of view of electronic warfare, one of the most important aspects of infrared is that it is a totally passive detection system. Unlike radar, infrared systems do not emit energy in order to perform detection. This characteristic provides considerable security during the observing period and complicates the problem of knowing when and where to jam infrared systems. A small satellite orbiting the Earth 500 to 1000 miles high would prove almost impossible to locate visually if its orbit were not known. Such a satellite equipped with an infrared detection system can provide a reasonably secure monitoring and reconnaissance system of infrared radiation activity from high-flying jets and large rocket launchings.

SOME EFFECTS OF SPACE OPERATIONS ON RADAR AND RADAR COUNTERMEASURES

Since radar has proven to be such a useful tool of electronic warfare, it will be informative to consider some applications of this tool in the context of space conflict. The fundamentals of radar will not be altered by operation in a space environment. However, to prove functional and realize its maximum usefulness some of the parameters will have to be extended by orders of magnitude. One of the principal differences between radar operations concerned with space system and other types is the great distances involved. Whether the radar is on the ground looking into space, or in space looking back at the ground, or looking into other parts of space, the distances are seldom less than a few hundred miles and may well extend to interplanetary ranges. To operate successfully over these great ranges, space radars require large average powers, large antennas, low receiver noise figures, and increased integration or data correlation times. Good estimates of the angular positions, range, and velocity of the target will also help to increase the probability of detection.

Adequate primary power for space-borne radar systems will present a serious problem for some time to come. Large areas of solar cells may be required to store electrical energy in batteries for intermittent high-power operation. A more attractive technique is the use of nuclear electrical power generators, especially when continuous operation is needed.

Large antennas may be used to help offset the high-power requirements. The lack of gravity and air forces simplifies design, but the antenna usually has to be folded in some manner (as discussed above) during the launch phase.

The advancement of receiver techniques will also enhance maximum radar ranges for space work. The use of masers or other cryogenic techniques to reduce the receiver noise temperature will prove very attractive. The effective sensitivity of the receiver can also be improved by increasing the signal integration time. This approach will require extensive use of coherent signals for transmission and for reference upon reception. The correlation principles outlined in Chapter 3 will prove very useful here.

In general, it can be seen that the successful use of radar for space missions primarily requires the extension and refinement of present techniques, rather than any fundamental variations. However, the applications of radar to space operations do introduce some interesting problems. One of the most vexing is the use of radar for ICBM tracking and identification.

ICBM's commence re-entry at about 22,000 fps, and satellites commonly have velocities in the order of 24,000 fps (see Fig. 8–10). This speed increases the range of Doppler frequencies to be searched. However, once detection has been effected, the velocity of the space vehicle is usually very predictable. Even during ICBM re-entry the deceleration follows a definite pattern that can be estimated in advance.

The gradual diminishing of the atmosphere from the ground to the vacuum of outer space affects radar operation. Between 30 and 300 miles ultraviolet sunlight produces ionospheric layers of free electrons that absorb and reflect radio signals. The maximum frequency that will be reflected depends upon sunspot activity, day or night, as well as the incident angle between the wave and the layer. Vertically, the maximum frequency reflected varies from approximately 2 to 12 megacycles, whereas beams directed toward the horizon may have to be as high as 30 to 70 megacycles to penetrate into space.

At radar frequencies the ionosphere produces very little signal attenuation, although it can rotate the plane of polarization of the wave because of the bi-refringent nature of the electron field combined with the Earth's magnetic field (Faraday Effect).

Unlike the ionosphere, the auroral activity over the Earth's magnetic poles will reflect radar signals. The effective cross section can be in the order of thousands of square meters.

Ordinary atmospheric absorption, which affects conventional (radar and target-inside atmosphere) long-range radars, especially those operating at higher radar frequencies, is less of a problem when the radar points into space.

If all the air was of constant density, it would extend to an altitude of about four nautical miles, that is, all the air straight up gives as much attenuation as would four miles horizontally. Of course, water vapor and rain are less prevalent at high altitudes. At 59,000 megacycles the oxygen resonance is at a maximum and the total attenuation (one-way) is about 100 db. Thus the atmosphere can be made to serve as a screen, preventing radiation from the surface from reaching space. This condition would give some privacy and immunity from ECM to inter-satellite communications systems, and would screen out ground clutter from radars searching for satellites or ICBM's.

Meteors reenter the atmosphere with speeds great enough to produce considerable ionization and their trails can be detected readily by radio or radar (even B-B-sized meteors can be seen). Their velocities are great enough so that they are seldom confused with possible ICBM's. The shock wave and ablation around ICBM nose cones result in an engulfing layer of ionization that can become opaque to radar signals during a portion of re-entry. This opacity would interrupt the jamming signals from any ECM equipment located in the nose cone. Also, it can alter the radar cross section of the nose cone, since the ionization could be totally reflecting or highly absorbent at various times, depending upon the electron concentration.

At altitudes beyond 100 to 200 miles the atmosphere offers almost no drag, and once in motion, light objects can follow the same trajectories as heavy ones. Thus two objects shaped like a satellite or an ICBM nose cone can appear identical to a radar, yet one may be a hollow metallized plastic balloon decoy weighing a few pounds whereas the other is real and weighs tons. It could be practical to make hundreds of such decoys, and it would be the radar's responsibility to identify the real object. Any object large and heavy enough to contain a warhead must be suspected of being a threat and techniques must be developed for establishing an object's physical characteristics. One of the most important identifying characteristics is the object's mass. The best measure of mass is acceleration resulting from some force, and acceleration can be detected as a change in velocity. Thus, radars can be used for decoy discrimination if they can be made sensitive enough to detect small velocity differences or changes in orbital periods. Satellites that pass through the lower altitudes have lifetimes limited by drag, providing a basis for discrimination by the deceleration that takes place.

Light balloon decoys of ICBM nose cones will not penetrate the atmosphere, and their deceleration can be noticed as high as 50 to 100 miles. Decoys that will re-enter far enough to give the impression of a serious threat must be more substantial. If the decoy has the same shape as the nose cone and is to have the same deceleration, it must have about the same weight, and might as well contain a warhead. If it is lighter it must be of a different shape to have similar deceleration, and usually must be smaller. Thus, size is a possible discriminatory criterion. Tumbling rates and stabilizing oscillations might also be used as identifying characteristics.

In addition to launching balloons or special decoys, a cloud of targets can be created by cutting the expended ICBM tankage into small fragments with high-explosive charges. A 1 fps separation velocity at engine burn-out will result in approximately one-mile separation at re-entry, so an extensive cloud of fragments requires only moderate velocity difference.

Nose-cone radar cross sections can be reduced by a covering of radar-absorptive material, or a metal shield shaped to have a low radar return in one direction (stabilized, non-tumbling nose cone). Such coverings might reduce radar ranges by a factor of 2 or 3 up to the time of re-entry, when they either would be released or would burn off.

Active ECM transmission from satellites is not effective against conventional radars searching for ground or airborne targets because the radar is normally much closer to the target than to the jammer. Also, the radar antenna discriminates against signals from other than the target direction. When the target is located in space it is so far from the radar that the jamming power required makes jamming more attractive.

However, since decoys require no electrical power, can be light, and do not have to be propelled in the absence of air, they are particularly attractive for space ECM applications.

Nuclear explosions can greatly affect the operation of radars in space. For conventional radars operating entirely within the atmosphere the most damaging effects are usually blast and thermal radiation. When human operators are with the radar, the necessity of their survival does not allow any damage to electronics in the vicinity. When the radar must propagate through the upper atmosphere, the normal nature of the ionosphere can be greatly disturbed by the energy from the nuclear bomb. Substantial increases in electron density can increase the critical frequency to a point above the radar bands, and make the ionosphere opaque and highly absorbent to radar. The duration and extent of the localized ionospheric enhancement depends upon the magnitude and altitude of the burst. If normal radar operation can be prevented for a matter of minutes, the delay could be sufficient to successfully defeat ICBM or satellite defense systems.

BIBLIOGRAPHY

CHAPTER 1
1. Newman, John R., *The World of Mathematics*, Vol. 4, Simon and Schuster New York, 1956.

CHAPTER 3
1. Goldman, Stanford, *Frequency Analysis, Modulation, and Noise*, McGraw-Hill Book Company, Inc., New York, 1948.
2. ————, *Information Theory*, Prentice-Hall, Inc., Englewood Cliffs, N. J., 1953.
3. Mood, Alexander M., *Introduction to the Theory of Statistics*, McGraw-Hill Book Company, Inc., New York, 1950.
4. Ridenour, Louis N., *Radar System Engineering*, McGraw-Hill Book Company, Inc., New York, 1947.
5. Shannon, Claude E. and Weaver, Warren, *The Mathematical Theory of Communication*, University of Illinois Press, Urbana, 1949.
6. Woodward, P. M., *Probability and Information Theory, With Applications to Radar*, McGraw-Hill Book Company, Inc., New York, 1955.

CHAPTER 4
1. Adams, Carsbie C., *Space Flight*, McGraw-Hill Book Company, Inc., New York, 1958.
2. Hudson, Ralph G., *The Engineers' Manual*, John Wiley and Sons, Inc., New York, 1939.
3. Moroney, M. J., *Facts From Figures*, Penguin Books Inc., Baltimore, Maryland, 1958.
4. Morse and Kimball, *Methods of Operations Research*, The Technology Press of MIT and John Wiley and Sons, Inc., New York, 1951.
5. Pierce, John R., *Electrons, Waves and Messages*, Hanover House, New York, 1956.

CHAPTER 5
1. Applebaum, Sidney and Howells, P. W., "Waveform Design for Tomorrow's Radars," *Space/Aeronautics*, October 1959, pp. 186–195.
2. Costas, J. P., "Poisson, Shannon, and the Radio Amateur," *Proceedings of the IRE*, December 1959, pp. 2058–2068.
3. Goldman, Stanford, *Frequency Analysis, Modulation and Noise*, McGraw-Hill Book Company, Inc., New York, 1948.
4. ————, "Some Fundamental Considerations Concerning Noise Reduction and Range in Radar and Communications," *Proceedings of the IRE*, May 1948, pp. 584–594.
5. Hall, W. M., "Prediction of Pulse Radar Performance," *Proceedings of the IRE*, February 1956, pp. 224–231.
6. Kaplan, S. M., "The Statistical Properties of Noise Applied to Radar Range Performance," *Proceedings of the IRE*, January 1951, pp. 56–60.

7. MIT Members of the Staff of the Radar School, *Principles of Radar*, McGraw-Hill Book Company, Inc., New York, 1946.
8. MIT Radiation Laboratory Series, *Radar System Engineering*, Vol. 1, Edited by Louis N. Ridenour, McGraw-Hill Book Company, Inc., New York, 1947.
9. MIT Radiation Laboratory Series, *Threshold Signals*, Vol. 24, Edited by James L. Lawson and George E. Uhlenbeck, McGraw-Hill Book Company, Inc., New York, 1950.
10. Sargent, R. S., "Moving Target Detection," *Electronics Magazine*, September 1954, pp. 138–141.
11. Woodward, P. M., *Probability and Information Theory, With Applications to Radar*, Pergamon Press. Published in U. S. by McGraw-Hill Book Company, Inc., New York, 1953.

CHAPTER 6

1. Hund, August, *Short-Wave Radiation Phenomena*, McGraw-Hill Book Company, Inc., New York, 1952.
2. Kraus, John D., *Antennas*, McGraw-Hill Book Company, Inc., New York, 1950.
3. Silver, Samuel, *Microwave Antenna Theory and Design*, MIT Radiation Laboratory Series, Vol. 12, McGraw-Hill Book Company, Inc., New York, 1949.
4. Summary Technical Report of NDRC, Vol. 3, *The Propagation of Radio Waves Through the Standard Atmosphere*, Washington, D. C., 1946.

CHAPTER 7

1. Bellman, Richard, *Dynamic Programming*, Princeton University Press, Princeton, N. J., 1957.
2. Bliss, G. A., *Lectures on the Calculus of Variations*, The University of Chicago Press, Chicago, 1946.
3. Feller, William, *An Introduction to Probability Theory and Its Applications*, John Wiley and Sons, Inc., New York, 1950.
4. Goldstein, Herbert, *Classical Mechanics*, Addison-Wesley Press, Inc., Cambridge, Mass., 1950.
5. Harvard Computation Laboratory, *Tables of The Cumulative Binomial Distribution*, Harvard University Press, 1955.
6. Luce, R. D. and Raiffa, Howard, *Games and Decisions*, John Wiley and Sons, Inc., New York, 1957.
7. McKinsey, J. C., *Introduction to the Theory of Games*, McGraw-Hill Book Company, Inc., New York, 1952.
8. Von Neumann, John and Morgenstern, Oskar, *Theory of Games and Economic Behavior*, Princeton University Press, Princeton, N. J., 1944.

CHAPTER 8

1. Buchheim, R. W., et al., "Some Aspects of Astronautics," *IRE Transactions on Military Electronics*, **2**, 1, December 1958, p. 8.
2. Greene, Jack, "Noise Factor and Noise Temperature," *Proceedings of the IRE*, January 1958, p. 2A.
3. Manoogian, Haig A., "The Challenge of Space," *Electronics*, **32**, 17, April 24, 1959, p. 65.

4. Pierce, J. R. and Kompfner, R., "Transoceanic Communication by Means of Satellites," *Proceedings of the IRE*, March 3, 1959, p. 372.
5. Stever, H. Gayford, "Our Interest in Space and Its Technology," *IRE Transactions on Military Electronics*, December 1958, p. 3.
6. Swerling, Peter, "Space Communications," *IRE Transactions on Military Electronics*, December 1958, p. 20.
7. Van Allen, James A., "Radiation Belts Around the Earth," *Scientific American*, **200**, 3, March 1959, p. 39.
8. Crupi, R. L., "Final Engineering Report on Electronic Techniques for Communication Satellites and Tracking Systems," Report No. AZP–129, Convair, Astronautics Division, San Diego, California, 1959.
9. Lund, W. W., Jr., "Some Design Consideration for Communication Satellites," Report No. AZM–060, Convair, Astronautics Division, San Diego, California, 1958.

INDEX

A

Absorbing materials, radar, 16
Absorption bands, 144
Action systems, 10–13
Active jamming, 16 (*see also* Jamming)
Active satellites, 188, 189, 192–193
Activity indicator, 53
Air-combat analysis, 17–21
Aircraft/missile guidance systems, 29
Air-defense system, 11–12, 29 (*see also* Defense systems)
Amplifier, IF, 85–89, 90, 92, 95, 105, 109, 112, 114, 124
Amplitude comparison monopulse, 148–149
Amplitude detector, 89–90, 91, 95, 99, 107, 112, 114
Amplitude probability distribution, 35, 43
Angle tracking, 111–114, 129, 131
Antenna systems:
 analysis, 132–153
 CW radar, 95, 96–97, 99–100, 102, 103
 directional, 50
 false targets, 128, 129
 gain (*see* Gain, antenna)
 noise jamming, 125, 126
 pattern (*see* Pattern, antenna)
 pulse radar range, 92–93
 radar, 84, 199
 satellite communications jamming, 192–193
 searching modes, 57, 60, 62–71
 space systems, 178, 180–182, 183, 193, 195–197, 199

Antenna systems (Continued):
 thermal noise, 85
 tracking, 111, 113, 114
A priori probabilities, 33
Atmospheric propagation, 139–143
Attenuation, 142, 143, 144
Autocorrelation, 40–42
Automatic gain control, 112, 114
Automatic tracking systems, 114, 130

B

Bandpass (*see* Passband)
Bands, radar, 82
Band width:
 antenna systems, 136, 137–139, 195
 CW radar, 95, 96–102
 global communication, 188, 192
 IF amplifier, 86–89
 pulse-Doppler radars, 106–108
 pulse radar, 94
 reconnaissance capability, 74, 75, 76, 79
 repeater-jammer, 129
 space communication, 182, 195
Band Width, 74, 75, 76, 79
Barometric fuze, 11
Barrage jamming, 6, 13–14, 18–21, 125, 127–128
Barrier curves, 17
Battery power, 187
Beaconing effect, 2, 6
Beam rotation, 147
Beam width:
 antenna diameter, 152–153
 gain, antenna, 137
 horizontal searching, 67, 70–71

206 INDEX

Beam width (Continued):
  search antennas, 150–151
  tracking antennas, 145
Blind ranges, 109
Blip-scan ratio, 56
Boltzmann's constant, 75, 85, 117, 180
Bomb-load weight, 7–9, 170–175
Broad-band barrage jamming, 13–14, 36
Brute-force methods, 13–14 (see also Noise jamming)
Burn-through range, 163

C

Central limit theorem, 35
Ceramics in space systems, 197
Chaff:
  basic theory, 15
  employment, 3
  operational considerations, 6
  radar jamming, 130–131
Chaff-corridor seeding, 15
Chemical batteries, 187
Circular polarization, 135–136
Clutter, radar, 94, 96, 98, 100, 101–102, 103, 108, 110, 131, 144
Coded modulation, 2
Coherent detection, 38–42, 45
Coherent integration, 91–92, 93, 99, 108
Communications systems:
  definition, 1
  design, 48–49
  directional antennas, 50
  ECCM fixes, 55
  global, 187–193
  intercepting, 54
  point-to-point, 27–28
  radar (see Radar)
  satellite relay stations, 176
  space application, 177, 178, 179–187
Confusion signals, 47
Conical-scan tracking system, 112–113, 116, 129–130, 146–147
Continuous variate, 33
Control systems, 50
Convair, 170
Correlation:
  long pulse detection, 49
  matched-filter processing, 122
  principles, 199
  signal detection, 39–42
Cosmic noise, 181

Cross-correlation technique, 39–41
Crossover range, 20, 22
Cross section, radar, 117
Curvature, Earth's, 177
CW radars, 94–103, 108, 109, 114, 116, 117–118, 121, 123, 124, 131

D

Damage function, target, 170–171
Data analysis:
  defense effectiveness, 175
  definition, 10–13
  electronic intelligence, 52–56
  reconnaissance satellite, 194
Deception jammers, 14–15
Decision systems, 10–13
Decoys:
  ICBM defense, 177
  optical system jamming, 194
  radar discrimination, 200–201
  technique, 15–16
Defense systems:
  air defense, 11–12, 29
  ECM effectiveness, 174–175
  effectiveness function, 163–169
  strategic deterrent system, 177–178
  tactics example, 21–25
  target damage probability, 170–174
Delay line system, 110, 111, 124
Deployment, 53–54
Design of ECM equipment, 154–169, 175
Detection:
  air-defense philosophy, 11–12
  CW radar, 95–96, 102–103
  Doppler filter, 97
  ECM presence, 42–50
  false targets, 129
  ICBM tracking, 199–201
  IF amplifier, 85–89
  incoherent and coherent, 38–42, 45
  infrared technology, 198
  low-flying attackers, 141
  Moving Target Indication, 110–111
  new techniques, 54–55
  problems, 2
  pulse radar, 29, 89–92
  radar transmission, 123–125, 126
  range equations, 116–117
  reconnaissance, 56–57, 60–61, 74, 194
  searching modes, 62–71

Detection (Continued):
  space systems, 179–187, 194
  tracking, 111–116
Detection Band Width, 74, 75, 80
Difference frequency, 85–89, 95, 119 (*see also* Intermediate Frequency signal)
Diffraction, 142
Dipoles, 196
Directional antennas, 50
Directivity, antenna, 136
Dispersion, 177–178
Distance Early Warning, 12
Doppler filters, 96–102, 106–108, 109, 116, 121, 124, 130
Doppler frequencies, 199
Doppler shift, 95, 96, 103, 105, 109–111, 186
Downward-searching mode, 57–58, 60–65, 71
Drone mission, 65
Duplexing, 84
Dwell Time, 75
Dynamic development of ECM tactic, 5–7
Dynamic programming, 163

E

Effectiveness function, 163–169
Electrical power in space, 186, 187 (*see also* Power systems)
Electromagnetic propagation, 132, 139–143
Electromagnetic radiation, 1–2
Electromagnetic waves, transit times, 185–186
Electronic Counter-Counter Measures:
  definition, 1
  design, 47
  frequency diversity, 192, 193
  profitability, 11
  protective devices, 48
  radar, 127
  reconnaissance relation, 4–5
  techniques, 55–56
Electronic Counter Measures:
  ECCM fixes, 55
  effectiveness function, 163–169
  general philosophy, 10–25
  global communications, 188, 191–193
  jamming (*see* Jamming)

Electronic Counter Measures (Cont.):
  noise, probability, and information recovery, 26, 42–50
  objectives, 154–155
  optimization of equipment characteristics, 155–163
  overview, 1–9
  payload distribution, 170–175
  radar, 81–82, 84, 111, 116, 123–131, 198–201
  space application, 182–183, 198–201
  tracking systems, 111, 113
Electronic noise, 29 (*see also* Noise)
Electronic Order of Battle, 53–54
Electronic reconnaissance, 51–80 (*see also* Reconnaissance)
Electronic warfare:
  antenna systems, 132–153
  basic problems, 1–9
  electronic reconnaissance, 51–80
  noise, probability, and information recovery, 26–50
  optimization, 154–175
  radar considerations, 81–131
  space era, 176–201
  technique and tactics, 10–25
Elliptical polarization, 135–136
Environmental tolerance, 179, 183
Epoxy resins, 197
Equipment-weight, 9, 170–175
Extraterrestrial bases, 178

F

False alarm, probability of, 43–45
False-target generator, 15
False targets, 128–131, 137
Faraday Effect, 200
Ferret receivers, 52, 54, 73–80, 194
Field-strength pattern, 135 (*see also* Pattern, antenna)
Filtering:
  CW radar, 95, 96–102
  long pulse, 49
  low pass, 37, 38
  Moving Target Indication, 110, 124
  passband, 37–38
  pulse-Doppler, 104–108
  radar filters, $87n$
  target detection, 119–120
  transmission spectra, 122

Fixes:
  classification, 127
  design, 47
  swept-spot jammer defense, 14
  techniques, 55
Flush antennas, 195–196
Fourier transform, 41–42
Frequencies (*see also* Band width):
  antenna band width, 137–139
  attenuation effects, 144
  diversity, 192, 193
  FM techniques, 98–101, 103, 105, 116, 120, 127
  IF signal, 84–89, 90, 95, 100–101, 104–105, 109–111, 112, 114, 116, 118–119, 120, 122, 124
  intercept identification, 72–73
  radar, 84–90, 95, 96, 98–102, 104–105, 107–108, 109–111, 113, 116, 118–119, 121, 127–128
  reconnaissance:
    receivers, 76–78, 79
    resolution, 74
  space:
    antennas, 195–196
    communication, 181, 188
Frequency modulation techniques, 98–101, 103, 105, 116, 120, 127

G

Gain, antenna, 136–137, 138, 143, 149, 151, 152, 153, 178, 180–182, 183, 193
Gate range tracking system, 114–116, 131
Guassian distribution function, 33–35, 36, 43–50
General tactic, 3, 5 (*see also* Tactics of electronic warfare)
Geographical resolution, 53, 58–59, 62–71
Gimbals, 145
Global communications, 187–193
Gravity, Earth's, 177
Ground-based antennas, 151–153

H

Hall, W. M., 19*n*, 21
Hardening, 177–178
Hedging against uncertainty, 164

High-speed transmissions, 2
Home-on jam, 16, 23, 163–164
Horizontal-plane pattern, 133 (*see also,* Pattern, antenna)
Horizontal searching mode, 58–59, 65–71, 149–151
Hostile environment, 6

I

Impedance matching, 138–139
Incoherent detection, 38–42
Indicators:
  radar, 90, 91, 95–96
  strategy, 53
Inflatable antennas, 196
Information recovery:
  ECM presence, 42–50
  effectiveness function, 163–169
  electronic intelligence, 52–56
  initial intercept, 62
  intercept correlation, 71–73
  noise environment, 26, 35–38
  reconnaissance, 51–80
  space detection, 194
  theory, 27–30
  tracking (*see* Tracking)
Infrared cells, 194
Infrared guidance system, 197
Infrared technology, 197–198
Integration:
  CW radar, 99
  pulse-Doppler radars, 108
  pulse radar, 89–92
Intelligence, electronic, 52–56, 62, 164
Interception:
  air-combat analysis, 17–21
  correlation, 71–73
  horizontal resolution, 65–71
  overflight, 62–65
  probability, 56–57, 60–61, 149–151
  tactics example, 23–25
Intercepts, 53, 58–73, 149–151
Intercontinental Ballistic Missile:
  decoys, 194
  defense, 177, 178
  development, 187
  indicator reconnaissance, 53
  infrared detection, 198
  radar tracking, 199–201
Interference, electronic, 1–3

Intermediate Frequency signal, 84–89, 90, 95, 100–101, 104–105, 109–111, 112, 114, 116, 118–119, 120, 122, 124
Inverse conical scan repeater, 14–15
Irreversible detection, 42n

J

Jamming:
  air-combat analysis, 17–21
  antenna structure, 141
  bomb-load weight ratio, 7–9
  burn-through range, 163–164
  detection probability relation, 44–47
  ducts, 143
  dynamic tactic development, 6–7
  ECM application, 1, 48
  global communications, 188, 191–193
  ground-based antennas, 151
  ICBM, 200, 201
  incoherent detection, 38
  infrared systems, 198
  optical systems, 194
  philosophy, 11–17
  problems, 2–3
  radar, 116, 123–131
  side lobes, 137
  signal information recovery, 35–50
  space communications, 182–183
  tracking antennas, 144

K

Kill probability 7–9, 23
Klystrons, 84, 85, 95

L

Lagrange multiplier, 160–162, 163
Launch barrier, 17
Launching paths, 184–185
L-band radar, 4
Left-hand elliptical polarization, 135
Lethal radius, 171
Linear detection, 42
Linear polarization, 135–136
Linear scanning, 60, 61
Local oscillator signal, 84–85, 95, 97, 100, 103–104, 105, 109, 116, 121, 122, 124

Look-through, 7
Lubricants in space, 197

M

Magnetron oscillator, 82–84, 95, 108, 127, 128
Manual tracking, 146
Masers, 199
Matched-filter radar, 122, 124, 129
Meteoroid protection, 183–184, 196, 200
Meteorological forecasting, 198
Micrometeorites, 196
Microwave radar, 198
Mission restrictions, 6
Mix, ECM-bomb, 171–172, 173, 174
Mixing radar signal, 84–85, 95, 100, 104, 105, 106, 110, 116, 120–121, 122, 124
Mobility, 177–178
Monopulse angle-tracking systems, 130, 147–149
Moon bases, 178
Moving Target Indication, 109–111, 124
Multiple-beam tracking, 149
Mylar, 197

N

Narrow-band fiilter, 95, 96, 119
Noise:
  cancellation technique, 30
  CW radar, 95, 96, 98, 99, 102–103
  ferret receiver sensitivity, 74–75
  frequency spectrum characteristics, 188
  information recovery relationship, 26
  jamming (see Noise jamming)
  Moving Target Indication, 110–111
  probability-distribution theory, 33–35
  pulse-Doppler radars, 105–107
  pulse radar, 85–86, 88, 89, 91–92
  range of radar, 117, 118, 119–120
  scanning receivers, 79–80
  signal information recovery, 35–50
  space communication, 180, 181–183
  temperature, 180, 181–182, 183, 199
  theory, 26–30
Noise figure, 85–86
Noise jamming:
  angle tracking, 114
  incoherent detection, 38

Noise jamming (Continued):
  philosophy, 13–14
  radar, 18–19, 124–126, 127–128, 129
  space communication, 183
  techniques, 44–47
Noiseless channel, 26–27
Non-coherent integration, 91–92, 99
Non-coherent pulse radar, 83
Non-linear detection, 42
Normal probability distribution function, 33–35
Nuclear explosions in space, 201
Nuclear power in space, 187, 199

O

Observation range, 177
Omnidirectional antenna, 192
Operation information, reconnaissance, 52–54
Optimization of ECM, 154–175
Orbital altitude, 190, 191, 193–194
Orbital bombardment systems, 176
Orbiting deterrent system, 178
Overflight, 57–58, 60–65, 71
Overpressure, 171

P

Parabolic antenna systems, 136, 145–146, 151–152
Paralleled filter receiver, 76–77
Passband:
  jamming effectiveness, 45, 47
  receiving-filter, 37–38
  scanning receiver, 79
  video receivers, 76
Passive jamming, 16 (*see also* Jamming)
Passive satellites, 188, 189, 191–192
Pattern, antenna, 133–136, 137, 140–141, 146
Payload distribution, 170–175
Phase comparison monopulse, 148
Photographic optical systems, 194
Photographic reconnaissance, 56
Photomultiplier, 127
Plane polarization, 135–136
Plan-Position Indication, 90, 95–96
Plastics for space systems, 197
Point-to-point communications systems, 27–28
Polarization of antennas, 135–136, 140, 145, 196–197

Polyethylene, 197
Polystyrene, 197
Power systems:
  antennas in space, 195
  impedance matching, 138–139
  pattern, 135 (*see also* Pattern, antenna)
  radar application, 82–84, 117, 199
  space, 178, 179, 180–181, 182, 186–187, 199
Practical noise, 40–41
Principal-plane vertical pattern, 133 (*see also* Pattern, antenna)
Probability theory:
  ECM presence, 42–50
  effectiveness function, 165–166, 167–168
  elementary relationships, 31–35
  intercept, 149–151
  killing, 7–9, 23
  noise theory, 30
  overflight, 60–62
  satellite communication, 191
  signals:
    identification, 72–73
    interception, 56–57
    recovery, 35–50
    redundant, 49
  space communication, 179–180
  successful mission, 172–174
  survival ratio, 8
  target:
    damage, 170–173
    detection, 117
Procurement of ECM equipment, 154, 175
Propagation, 132, 139–143
Pseudo-noise generation, 2
Pulling, antenna, 138
Pulse coding radars, 129
Pulse-compression radar system, 124, 129
Pulse-Doppler radars, 89, 103–111, 116, 117–118, 121, 124, 131 (*see also* Radar)
Pulse integration, 19–21
Pulse radar, 27, 29–30, 82–94, 103, 108, 109, 114, 116, 117–118, 122, 123 (*see also* Radar)

R

Radar:
  absorbing materials, 16
  air-combat analysis, 17–21

INDEX   211

Radar (Continued):
   air-defense system, 11–13
   analysis, 81–131
   band width, 138
   chaff, 15
   deception jammers, 14–15
   decoy, 15–16
   design, 48–49
   directional antennas, 50
   ECCM fixes, 55
   ECM advantage, 48
   home-on jam, 16
   information recovery, 27–30
   infrared application, 198
   intercept probability, 149–151
   noise jamming, 13–14
   parabolic dishes, 145
   power pattern, 135
   propagation over Earth, 140–141
   satellite antenna, 195
   space application, 194, 198–201
   stand-off jamming, 17
   submarine searching, 4
   tracking antennas, 144–149
   van mobility, 60
Radar-barrier, 17
Radar Order of Battle, 53–54
Radiation:
   antenna materials, 197
   detectors, 194
   electromagnetic, 1–2
   pattern, 133–136, 137 (see also Pattern, antenna)
   protection, 183, 184–185
   radar operation, 201
Radio-control system, 48–49
Radio Frequency energy, 2
Radio power tubes, 127
Radio wave propagation, 142–143
Rand Corporation, 170
Random noise, 30, 33–35, 39, 40, 41, 42–50
Range gate, 14, 124, 131
Range-gate pulse, 121, 122
Range-gate stealer, 14
Ranges, radar:
   CW radar, 102–103
   detection equations, 116–117, 120
   MTI, 110–111
   pulse-Doppler radar, 109
   pulse radar, 92–94
Range tracking, 114–116, 130

Receivers:
   modes, 97
   reconnaissance, 73–80
   sensitivity, 74–75, 76, 78, 79, 80
Receiving principle, 148–149
Reciprocity Principle, 133–135
Reconnaissance:
   antenna structure, 138, 141
   confusion signals, 47
   ECM effectiveness function, 164, 165
   electronic analysis, 51–80
   function, 4–5
   ground-based antennas, 151
   infrared application, 198
   jamming transmitters, 2
   parameters of radar signal, 48
   space, 177, 178, 193–194, 195
   tactic weight, 6
Redundancy, 49, 50
Repeater-jammers, 129
Rectangular pulse, 87
Right-hand elliptical polarization, 135

S

Saddle points, 169
Salvoes, tactics example, 23–25
Satellite systems:
   antennas, 195–197
   bases, 178
   ECM transmission, 201
   global communications, 188–193
   infrared technology, 197–198
   launching paths, 185
   reconnaissance, 54–55, 193–194
   relay stations, 176, 177
   tracking antennas, 151
   warning systems, 176, 177
Saturation signals, 47
S-band radar, 4
Scanning:
   antennas, 145, 146–147, 149–151
   CW radar, 97
   modes, 60, 61, 62–71, 75
   radar tracking, 111, 112–113
   receiver, 78–80
Scattering, 142–143, 144
Search antennas, 149–151
Searching modes, 57–71, 149–151
Security problems, 2
Sensitivity, receiver, 85–86, 96
Sidebands, radar, 97–99, 100
Side lobes, 136, 137, 144, 145

Signals:
  atmospheric effect, 199–200
  attenuation, 144
  coding, 193
  definition, 27
  detection (*see* Detection)
  electronic reconnaissance, 56
  extraction techniques, 49–50
  information recovery, 26–30 (*see also* Information recovery)
  interception and detection, 56–57
  jamming (*see* Jamming)
  propagation effect, 139–140, 142–143
  radar, 84–92, 95, 96, 99, 100, 102–103, 105–107, 110–111, 119–121, 122, 125–126, 128
  reconnaissance receiver modes, 76–80
  recovery, 35–50
  space communication, 185–186
  tracking, 111–116
Signal-to-noise ratio, 56, 88, 89, 91, 92–93, 102–103, 107, 110–111, 117, 119–120, 122, 125–126, 144
Signatures, radar design, 48
Simultaneous Lobe Comparison, 113–114
Sniperscope, 198
Solar energy in space, 187, 199
Space era:
  antennas, 151
  electronic warfare analysis, 176–201
Spectral distribution, 35, 41–42
Spectrum, radar, 118, 120, 121, 122, 123
Spike autocorrelation, 41–42
Spiral antennas, 195
Split-gate range tracking, 114–116
Spot jamming, 6, 13–14, 128
Sputnik I, 196
Stabilizing oscillations, 201
Stand-off jamming, 17
Stanford Research Institute, 170
Statistics:
  correlation, 39–42
  probability theory, 30, 31–35 (*see also* Probability theory)
Steerable antenna, 192
Strategic Air Command, 3, 188
Strategic deterrent system, 177–178
Strategic planning, 51–52
Submarine searching, 3–4
Superheterodyne receiver, 78
Surface-to-air missile system, 23
Survival specifications, 179
Swept filter, 97
Swept-spot jammer, 14

T

Tactics of electronic warfare:
  basic philosophy, 10–13
  dynamic development, 5–7
  example, 21–25
  overflight resolution, 62–65
  reconnaissance, 51, 53–54
  techniques, 13–21
Targets:
  damage function, 170–173
  detection, 117–121, 125
  tracking systems, 111, 146–149
Technical reconnaissance data, 54–56
Technical supremacy, 3–5
Television, 194
Temperature:
  noise, 180, 181–182, 183, 199
  space stabilization, 183
Terrestrial noise, 181
Terrestrial warfare, 176, 177–178
Thermal diode, 127
Thermionic converter, 187
Tracking:
  antennas, 144–149
  ICBM identification, 199–201
  infrared systems, 198
  jamming, 129–131
  radar systems, 111–116
  space communication, 186
Transistors, 184
Transmitted power (*see also* Power systems):
  pulse radar range, 93
  space environment, 178, 179, 180–181, 182
Traveling wave tubes, 127
Tumbling rates, 201
Tuned circuit problem, 80
Tunnel diodes, 184

U

Uncertainty hedging, 164
Unfolding antennas, 196

V

Vacuum tubes, 84, 184, 187
Van Allen radiation belts, 184–185

INDEX   213

Velocity tracking system, 115, 116, 130
Vertical-plane pattern, 133 (*see also* Pattern, antenna)
Video amplifier, 90
Video receivers, 76
Voltage Standing Wave Ratio, 138–139
Volume constraint of payload, 155–156, 158–163, 167

W

Warfare, electronic (*see* Electronic warfare)
Wavelength, radar, 93, 95
Weapons systems:
  ECM tactic development, 5–7

Weapons systems (Continued):
  jamming (*see* Jamming)
  payload distribution, 170–175
  space planning, 176
  technical supremacy, 3–5
Weight constraint of payload, 7–9, 155–163, 167, 170–175
White noise, 36, 37, 41, 127
White spectral distribution, 35
Wide-band barrage jamming, 127, 128
Wide-band noise devices, 127
Wide-open receiver, 76

Y

Yield, payload, 171

Synopses of Reprint Classics by Peninsula Publishing

Books can be ordered by mail or telephone.
See last page for instructions.

Ambient Noise in the Sea — Ambient noise, a subset of the ocean's total underwater noise, is defined as the residual, unwanted noise after all other noise sources have been identified. *Ambient Noise in the Sea* by Robert Urick encapsulates the body of knowledge on this subject. It discusses the sources of ambient noise, its variability, and its dependence on receiver depth, directionality, and coherence. This book is a *must* for engineers in the field of active and passive sonars, underwater sensor and weapons systems, and underwater signal processing.

Hardcover. 205 pages. Order Book No. P114. $29.95 USA, $31.95 Int'l.

An Introduction to Statistical Communication Theory — Written by David Middleton, pioneer in statistical communication theory, this classic established a unified approach to the basic theory and applications of random signals in communication systems. The book provides a detailed account of systems and their elements as operations and changes on signal and noise ensembles; it addresses the adaptation of statistical *decision theory* to communication problems. The book emphasizes system optimization and evaluation of threshold detection and extraction . . . system design . . . comparison between theoretical optimum and actual suboptimum systems . . . and structure of optimum systems in terms of existing elements.

Hardcover. 1100 pages. Order Book No. P107. $59.95 USA, $61.95 Int'l.

Digital Communications with Space Applications — This book defined an entire new technology for space communications. It was known, irreverently, as "The Bible" at the Jet Propulsion Laboratory. Subjects addressed include: C^3, data telemetry, tracking and ranging, coding, sequences and synchronization techniques. The authors, Solomon W. Golomb, Andrew J. Viterbi, Leonard Baumert, Mahlon Easterling and Jack Stiffer are among the foremost experts in the communications field.

Hardcover. 210 pages. Order Book No. P109. $19.95 USA, $20.95 Int'l.

ECM and ECCM Techniques for Digital Communication Systems — Ray H. Pettit. This release presents an overview of contemporary concepts and techniques in the area of ECM and ECCM for digital communications. Focuses on applicable models and procedures, and gains and losses in systems performance, graphic illustrations and practical examples. Written in concise survey style for engineering and management professionals.

Hardcover. 178 pages. Order Book No. P105 $21.95 USA, $23.95 Int'l.

Electronic Countermeasures — Originally published as a secret reference in the 1960s by the U.S. Army Signal Corps, the book is now declassified. Its 1100 pages cover signal intercept, jamming and deception fundamentals that are as valid today as when first written. Other subjects include intercept probabilities, receiver parameters, detection and analysis, direction finding, jamming technique, IR and acoustic countermeasures. More than 600 references and a list of authors that reads like Who's Who in Electronic Warfare.

Hardcover. 1100 pages. Order Book No. P103. $40.00 USA, $42.00 Int'l.

Enter my order as follows:

Book No.	Quantity	Unit Price	Total Cost
_____	_____	_____	_____
_____	_____	_____	_____
_____	_____	_____	_____
_____	_____	_____	_____
_____	_____	_____	_____
_____	_____	_____	_____

Total of Book Purchases _____

Calif. Residents add 7% sales tax _____

Postage & handling @ $1.50 per book _____

GRAND TOTAL OF ORDER   US$ _____

Charge to: ☐ Mastercard   ☐ Visa

Account Number: _____

Expiration Date: _____

_____
signature as it appears on charge card

_____
name (please print)

_____
organization (if required for shipment)

_____
address

_____
city   state   zip

_____

Air Mail Service (surcharges per copy):
US priority mail, $3.00
Canada, $5.00
Europe, $12.50
Asia/Africa/Oceana, $15.00
South & Central America, $6.00

Peninsula Publishing
PO Box 867
Los Altos, CA 94022 USA

Enter my order as follows:

Book No.	Quantity	Unit Price	Total Cost
_____	_____	_____	_____
_____	_____	_____	_____
_____	_____	_____	_____
_____	_____	_____	_____
_____	_____	_____	_____
_____	_____	_____	_____

Total of Book Purchases _____

Calif. Residents add 7% sales tax _____

Postage & handling @ $1.50 per book _____

GRAND TOTAL OF ORDER   US$ _____

Charge to: ☐ Mastercard   ☐ Visa

Account Number: _____

Expiration Date: _____

_____
signature as it appears on charge card

_____
name (please print)

_____
organization (if required for shipment)

_____
address

_____
city   state   zip

_____

Air Mail Service (surcharges per copy):
US priority mail, $3.00
Canada, $5.00
Europe, $12.50
Asia/Africa/Oceana, $15.00
South & Central America, $6.00

Peninsula Publishing
PO Box 867
Los Altos, CA 94022 USA

BUSINESS REPLY MAIL
First Class   PERMIT No. 722   Los Altos, CA 94022

POSTAGE WILL BE PAID BY ADDRESSEE

PENINSULA PUBLISHING
P.O. Box 867
Los Altos, California 94022

BUSINESS REPLY MAIL
First Class   PERMIT No. 722   Los Altos, CA 94022

POSTAGE WILL BE PAID BY ADDRESSEE

PENINSULA PUBLISHING
P.O. Box 867
Los Altos, California 94022